LEARNING
from the
OCTOPUS

LEARNING from the OCTOPUS

How Secrets from Nature Can Help Us Fight Terrorist Attacks, Natural Disasters, and Disease

Rafe Sagarin

BASIC
BOOKS
A Member of the Perseus Books Group
New York

Illustrations by Rafe Sagarin

Published by Basic Books,
A Member of the Perseus Books Group

Books published by Basic Books are available at special discounts for bulk purchases in the United States by corporations, institutions, and other organizations. For more information, please contact the Special Markets Department at the Perseus Books Group, 2300 Chestnut Street, Suite 200, Philadelphia, PA 19103, or call (800) 810-4145, ext. 5000, or e-mail special.markets@perseusbooks.com.

Designed by Linda Mark
Set in 12.5 Dante by the Perseus Books Group

Library of Congress Cataloging-in-Publication Data
Sagarin, Rafe.
Learning from the octopus : how secrets from nature can help us fight terrorist attacks, natural disasters, and disease / Rafe Sagarin.
p. cm.
ISBN 978-0-465-02183-3 (hardback) — ISBN 978-0-465-02981-5 (e-book)
1. Terrorism—Prevention. 2. Emergency management. 3. Ecology—Study and teaching. 4. Evolution—Study and teaching. I. Title.
HV6431.S224 2012
363.34—dc23
2011048842

This book is dedicated to Paula Loyd,
one of the most adaptable human beings I have ever known.

CONTENTS

FOREWORD

BIG IDEAS RARELY come from predictable sources. This is particularly true in the realm of national security.

When the U.S. national security state was created in 1947, our government made a large standing military permanent, created an international clandestine intelligence capability that continues to grow, and established a priesthood of national security "experts." Like all priesthoods throughout human history, the national security priesthood was open only to those who possessed vital secrets.

Secrecy has two functions: it establishes a fraternity of those on the inside, and it locks out those on the outside. Thus, if you are not a member of the fraternity or the priesthood, you are not welcomed into national security deliberations. Over time, closed systems produce fewer and fewer innovations, because closed systems, by definition, are based on certain increasingly unchallengeable fundamental principles.

The national security priesthood participates in established organizations and communicates through established publications such as *Foreign Affairs*. Some time ago a prominent academic reviewing stale foreign policy books in the *New York Times* lamented the absence of creative thinking in foreign policy circles. Predictably, she had defined those circles as Boston, New York, and Washington, in part perhaps because she taught at a leading Ivy League university.

There is not one chance in a thousand that anyone within that narrow (in every sense of the word) priesthood would have thought to apply elemental principles of biology to a fresh understanding of conflict in the new age of the twenty-first century as Rafe Sagarin has done in this innovative book. He is, praise be, not a member of the established national security priesthood, at least for the time being. But do not be surprised if, in coming months and years, you see ideas contained in *Learning from the Octopus* sprouting up—with or without attribution—that are premised on principles of adaptability and organic thinking.

Combining genius and common sense, Sagarin sees in the tide pools of Monterey, California, living organisms with much to teach us about twenty-first-century conflict. His timing is impeccable. For we are departing from a three-and-a-half-century period following the Peace of Westphalia in 1647 when the nation-state defined and conducted warfare. That warfare involved great uniformed national armies meeting in the field, exchanging men and materiel, until a white flag was raised by one side and victory was declared by the other. For, during this period, the nation-state possessed a monopoly on violence granted by the bargain between the state (government) and the nation (the people) that the state would guarantee the security of the nation if the people would declare their loyalty to the state.

Beginning in Vietnam, then into Afghanistan (first for the Soviets, and now for us), and into bitter days in Iraq, things began to change. Combatants didn't wear uniforms. They carried neither banner nor flag. They attacked civilian targets, in violation of the scripted "rules of war" contained in the various Geneva Conventions. And they behaved more like eleventh-century assassins than honorable regiments so familiar to the twentieth century.

The Westphalian era of conflict ended when the most powerful nation in human history could no longer guarantee the security of its citizens. That was on September 11, 2001.

Because U.S. forces invaded Afghanistan and Iraq expecting to meet more traditional enemies, they were caught off guard. And even as senior officers persisted in trying to conduct traditional military operations in an unconventional, irregular conflict, our troops found themselves forced to improvise. They had been trained and equipped for one kind of conflict and quickly found themselves in another. Massive multi-ton combat vehicles could not negotiate treacherous mountain passes and were overnight sitting ducks for IEDs (improvised explosive devices). The key word here is "improvised."

There has been, in fact, what had been called, in theory, "a revolution in military affairs." But contrary to those in the national security priesthood who invented this phrase, the revolution was not in highly sophisticated, highly technical, computer-directed super-systems. The revolution was in a return to a gritty house-by-house, virtually hand-to-hand combat in very close quarters. The revolution was also in winning the hearts and minds of tribal members, not in the rather graphic way Lyndon Johnson described, but often with a wad of cash.

The two exceptions to this transformation are in drones and robots. The drones, magically, are controlled halfway around the world. But they still need special forces and on-the-ground intelligence collection to know what primitive dwelling to target. The robots are beginning to prove their worth also at an elementary level in bomb detection. Thus, the vaunted technology revolution in warfare is operating much more at the grassroots combat level than at the geopolitical top-down level so preferred by the national security priesthood.

As Sagarin colorfully points out, there are precedents for this kind of adaptability in the natural world, and we—particularly our military officials and policy makers—should learn and take note. Leaving the stale "debate" over evolution to backward-looking political figures, Sagarin shows that biological science demonstrates how fauna and some flora adapt to changing conditions not necessarily in a process of seeking perfection. Nature is pragmatic. It has that in common with the best militaries of the past. Creatures change not necessarily to make themselves more beautiful or more exquisite. They do so because it helps them survive in a constantly changing environment.

For those of us who have studied military history and theory for long periods, the application of biological principles to human security is fresh, challenging, and exciting. It is a big idea in a realm where they are especially scarce.

Let us leave it to the father of all strategic thought, Sun Tzu, to make the point: "A military force has no constant formation, water has no constant shape: the ability to gain victory by changing and adapting according to the opponent is called genius."

The frustration of the American people with the two long wars in Afghanistan and Iraq is in large part because we now know that our opponents are not going to raise a white flag and sue for peace while we celebrate a great victory and sail triumphantly home. Our massive military superiority, even against the feared Soviet Union, has not brought indigenous rebels to their knees in either country. Something is wrong here.

The lesson is clear: genius, imagination, and adaptation must now replace raw power. The U.S. armed forces in the twenty-first century must adapt to the new/old low-intensity, irregular, unconventional conflict of today and tomorrow. Big army divisions, nuclear carrier task groups, and long-range bomber wings now control the battlefield and dictate the outcome of conflict very much less than in the twentieth century. So long as we borrow a trillion dollars from the Chinese and purchase their products, there is virtually no chance of an all-out nation-state war with them. Both of us are smarter than that and have much more to lose than to gain.

But conflict is a constant. And the lessons of the long wars already teach us that we will need to adapt. We will need smaller combat units, more special forces, much more sophisticated on-the-ground intelligence, drones and robots, bundles of cash for tribal chieftains, and, most of all, the ability to adapt to constantly changing circumstances.

Sagarin has precedent for his principle of natural security in none other than Niccolo Machiavelli, the sage of Florence, who thought a prince had to adopt certain attributes from nature's creatures, most particularly the lion and the fox. "It is necessary to be a fox to discover the snares and a lion to terrify the wolves." But Machiavelli was more perceptive than most in his understanding that a reformer, including a biologically oriented one, faces enormous institutional resistance, resistance to adaptation and against structures erected by national security establishments and priesthoods decades ago:

> There is nothing more difficult to take in hand, more perilous to conduct, or more uncertain of its success, than to take the lead in the introduction of a new order of things. Because the innovator has for enemies all those who have done well under the old conditions, and lukewarm defenders in those who may do well under the new. This coolness arises partly from

> fear of the opponents . . . and partly from the incredulity of men, who do not readily believe in new things until they have had a long experience of them.

Ironically, it is often the case that the "lukewarm defenders" transform themselves into red-hot advocates sometime after the principle of biological adaptation as applied to military affairs has become a well-established organizing principle. Such is human nature. Nature's creatures, as Sagarin shows, simply get on with the business of adaptation without fear or favor.

It is well worth contemplating the implications of an "organic" military, one that senses in its very being, especially at the combat level; how the threat shifts and changes; what new, often crude adjustments the opponent makes; how a new weapon becomes quickly neutralized by a simpler counter-measure; how the combat "fish" navigate the waters of their society; and then a military that acts quickly enough to cut across the opponent's cycle of change to take the initiative. This loop, invented by the former combat pilot John Boyd, became the centerpiece of military reform in the late twentieth century.

It will be interesting to observe the reception Dr Sagarin's big idea receives in established national security circles. Experience suggests that the real Establishment will ignore it or dismiss it. Younger, newer thinkers will discuss and debate it. But, most importantly, it will be amazing if young military officers and Afghan and Iraq combat veterans do not read, circulate, and vigorously discuss this thesis. And that is the circle that must be reached if there is to be hope for real national security reform anytime soon.

As a veteran military reformer, I was often reminded that no major military establishment in history had reformed itself absent a major military defeat. Military structures are understandably conservative, and warfare is too hazardous to shake things up.

That is, unless you are forced to. Sagarin has provided plenty of evidence from our current conflicts to show that the troops, the veterans of the kind of conflict that prevailed before Westphalia, understand the need for adaptation. Whether they have to be trained biologists to understand the natural principles for this is beside the point.

All those who care about the security of our nation, and of our children's generation, whether in uniform or not, should study and discuss Sagarin's argument. It offers a refreshing new way of thinking about security in a precarious and different new era and century. It shatters old modes of thinking in a constructive and challenging new way. It has profound implications for how we defend ourselves and preempt attack and aggression. And it perceptively realizes that security is no longer a purely military undertaking.

Learning from the Octopus is the way Nature would behave if she were in charge of the Pentagon and our national security.

Gary Hart
Kittredge, Colorado
U.S. Senator (Ret.); former member, Senate Armed Services committee; chair, American Security Project; co-chair, U.S. Commission on National Security/21st Century

PROLOGUE: UNNATURAL DISASTERS

ON THE MORNING OF DECEMBER 26, 2004, animals across Asia and Africa were acting strangely. Elephants elicited horrific bellows, herds of oxen bolted for higher ground, and domestic dogs refused to go on their morning walks along the beach. In some cases, bewildered humans followed the lead of their charges to higher ground, but many did not. Less than an hour later, the ocean was sucked back far from shore and a huge tsunami thundered all across India, Africa, and southern Asia, killing 225,000 people—one of the worst catastrophes in modern history.[1]

After the floodwaters retreated, international aid poured in, with particular attention paid to installing state-of-the-art tsunami warning systems across the region. Yet in comparison to the animal-based warning systems, these high-tech solutions are still fairly primitive. Just a few years after the tsunami, villagers in the Aceh province of Indonesia, one of the hardest hit areas, angrily stoned their tsunami alarm until it was destroyed. The villagers felt the annoyance of the system's false alarms outweighed its purported benefits in early warning.[2]

Destroying alarm systems that are supposed to protect us isn't uncommon. In the United States, residents of over 21 million households have tampered with, destroyed, or disabled their own smoke detectors because of the nuisance of false alarms.[3] In fairness to the makers of smoke and tsunami alarms, such technologies have only been around for a few decades—a fleeting fraction of Earth's long and violent experience with tsunamis, floods, and fire. By contrast, the surprisingly accurate security systems demonstrated by the animals before the tsunami have been developed and fine-tuned over billions of years, and this illustrates a major point: there is no technological solution that can prepare us for the risks of a highly variable and unpredictable world as well as the ancient natural process of adaptation.

Indeed, just a few weeks before the 2004 tsunami, the most technically sophisticated military force in the world inadvertently and quite publicly demonstrated how poorly adapted it was to its latest challenge. It happened during a pretty standard piece of military propaganda set up for the evening news. The U.S. secretary of defense was to helicopter in to the edge of a war zone to bolster the troops' morale, listen sincerely to their concerns, and assure them that all of America was fighting right there alongside them. But it didn't turn out that way for Secretary of Defense Donald Rumsfeld in Kuwait on December 8, 2004. To the cheers of several thousand soldiers assembled, Specialist Thomas Wilson, a 31-year-old Tennessee National Guardsman, pointedly asked the secretary why he and his fellow soldiers were being forced to rummage through garbage dumps to find armor to strap on to their vehicles, which provided inadequate protection in the combat zone. Rumsfeld was initially taken aback, then tartly retorted, "You go to war with the Army you have."[4]

It was a pivotal moment in how George W. Bush's war in Iraq was going to be interpreted. The left seized on it as yet another example of the "chicken hawks" in the Bush administration cruelly sacrificing their pawns in an elective game of geopolitical chess. The right amplified reports that Wilson was "fed" his question by a *Chattanooga Free Press* reporter[5]— more evidence that the liberal media was out to sabotage an essential front in the "War on Terror." Even fairly level-headed commentators couldn't help but contrast the scrappy U.S. soldiers rummaging around junk piles for "hillbilly armor" to weld to their vehicles against the disinterested, out-of-touch button-down government bureaucrat. But the most important contrast the exchange belied hasn't been well noted—it was the difference in *adaptability* demonstrated between soldiers like Specialist Wilson and a large security organization like the Department of Defense (DoD). This is, in fact, the same difference in adaptability between animals sensing and responding to a tsunami and tsunami alarms sensing and responding to something that may or may not be a tsunami.

For armies fighting a war, for health practitioners trying to ward off a flu pandemic, for first responders containing the damage from a natural disaster, for IT managers trying to protect a computer network, for resource managers trying to plan for a world dramatically altered by climate change, for CEOs worried about the next stock market crash, and for any citizen worried about the effects of any of these potential threats, adaptability is essential. If we want to interact with the world at all (and in a world of 7 billion people, we don't have much choice), having the ability to *change* how we interact with it is the only way we can survive.

For the troops on the ground, the process of adapting began soon after the invasion of Bagdad. They "went to war with the Army they had" (to paraphrase Rumsfeld), and it worked brilliantly for a while. With massive firepower, better training, and air superiority, even the most feared of Saddam Hussein's forces virtually collapsed in front of the advancing coalition. But as the old regime collapsed, the ground became rich for any number of new threats to sprout up. The threat environment radically changed.

Suddenly, thousands of soldiers, independently as individuals and linked through the units they fought with, were observing that hidden improvised explosive devices (IEDs) were becoming their biggest threat to security. Whereas the DoD had planned for a war against AK-47s, Scud missiles, and weapons of mass destruction, soldiers on the ground began to see their enemies as random trash piles, sudden fender benders in downtown traffic, and cell phones; hiding, distracting from, and detonating IEDs. By the time Wilson was so incensed as to dare breach military protocol to give a superior officer a dressing down, 266 of his colleagues had been killed due to IEDs.[6]

The soldiers adapted the best that they could—welding metal plates to their vehicles, blocking up culverts to eliminate the most obvious niches for bombers to use, and learning to identify the signs of hidden bombs in otherwise unremarkable debris. But their ability to adapt was limited by forces beyond their control—by the equipment they were given, by the available scrap metal, by the rules of engagement that they were ethically and legally bound by—and the casualties mounted.

By contrast, the Department of Defense had virtually unlimited resources, especially after the terrorist attacks of September 11, 2001, when no politically minded senator or representative would ever turn down a military appropriation request. What the DoD lacked was adaptability. Even as Specialist Wilson and his comrades were frantically tracking the rapidly changing tactics of insurgents, the DoD was slowly churning away on weapons systems and fighting procedures that had been dreamed up long ago in places far, far away from the streets of Baghdad and Fallujah. Rumsfeld's retort to Wilson belied a centralized view whereby small numbers of intellectuals design a battle plan and the accompanying technology years in advance, *and that's what you go to war with.* Moreover, even to bring the idealized technological solutions to deal with the threats theorized by DoD experts, the department was bound by a ponderous top-down procurement system in which a small number of large contractors submitted bids for development of weapons systems that inevitably ran over-budget and beyond the estimated timeline.

As a result, until Specialist Wilson's outburst, the upper reaches of the DoD were neither sensing changes in the threat environment nor responding quickly when new threats were correctly identified. And even after congressional outrage from the exchange between Wilson and Rumsfeld fueled calls to speed up production and deployment of mine-resistant ambush-protected vehicles (MRAPs), they did not arrive in Iraq in until November 2007, nearly three years later. By that time, an additional 1,589 of Wilson's colleagues had been killed due to IED attacks. The DoD solution certainly arrived too late to save their lives, but also too late to even deal with the original threat. A rapid downward trend in IED attacks and deaths was already well on its way by the time the MRAPs arrived in Iraq (see Figure 1, page xxii).

Nonetheless, DoD was able to claim that the MRAPs were ready just in time for a renewed offensive in the long-simmering war in Afghanistan. But the environment of Afghanistan was very different, much more rugged, than that of Iraq. Most of it was downright impassable to 14–24-ton vehicles like the MRAPs, which meant that the Taliban's cheap, beat up old Toyota pickup trucks (probably the most adaptable vehicle ever built) could operate at will without interference from the lumbering U.S. forces. The few roads in Afghanistan that were MRAP accessible quickly became targets for IED attacks (which had been only a minimal threat up until this point), so that simple travel to a meeting with local leaders became a cumbersome and dangerous affair, sometimes taking all day to move 12 kilometers or so.[7]

The military recognizes now how poorly adapted the first MRAPs are. In fact, after two years of deployment, nearly half of the 16,000 MRAPs produced are being put on "inactive status"[8] (kind of like your neighbor's old Camaro on blocks, only the MRAPs being put on blocks are only two years old and cost half a million dollars each). Even as soldiers were stuffing themselves into the brand-new MRAPs, contracts went out for the next generation of MRAPs—known as the MRAP-All-Terrain Vehicle (M-ATV) and weighing only 7–10 tons (a typical four-wheel-drive Toyota pickup weighs just over 1 ton).

It wasn't due to a lack of effort or resources that the NGOs responding to the tsunami and the Department of Defense failed so

Figure 1: Number of deaths per month of U.S. forces due to IEDs in Iraq (black) and Afghanistan (gray)

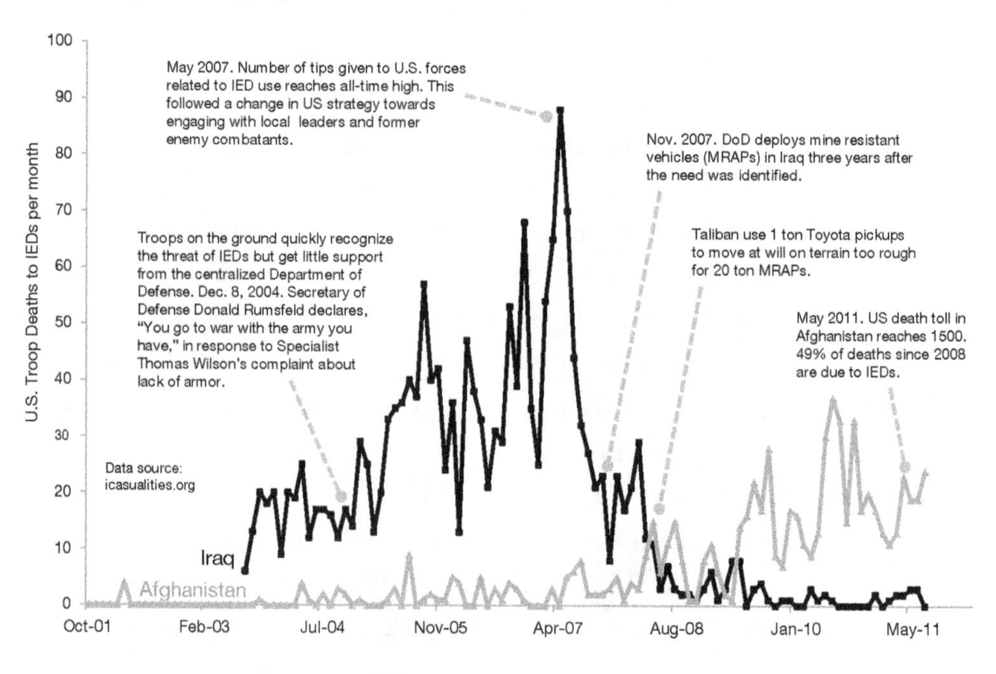

miserably to adapt relative to the lowly animals on the shore or the lowly grunt soldiers in their Humvees. In reaction to the disasters in Asia and in Iraq, trillions of dollars in aid and military budget have been spent trying to plan for, predict, and perfect our responses. But organisms in nature and soldiers on the ground don't have the luxury of limitless resources and hours of meetings and PowerPoint presentations to map out their future. In short, natural adaptive systems don't plan, they don't predict, and they don't perfect.

There is no scientific evidence to support the notion that evolved biological systems are "intelligently designed" or planned in advance. There are numerous natural examples to illustrate this point, but one look at the *Mola mola*—a huge slow-swimming fish so bizarre you would swear that you were looking at a fish that had its back half bitten off by a shark and then was rolled over with an underwater steamroller—will convince you that it followed no plan in its development as a fish. Yet in its own way of surviving for millions of years, it has been as wildly successful as a tuna or a trout. An unlikely creature such as the *Mola mola* emerged because the process of evolution doesn't tend toward any endpoint. It doesn't try to make an eye or an immune system or a beautiful fish. Evolution proceeds by solving survival problems as they arise. Many systems in society, by contrast, are littered with meticulously planned designs—the Maginot Line comes to mind—that were entirely unable to solve emerging threats from the environment.

These unplanned, evolved, adaptable organisms themselves don't make predictions. Why not? Simply because the complex world of continually changing and interacting biological organisms acting within a dynamic and networked matrix of biogeochemical stocks and flows that they live in is not predictable. At best, organisms anticipate events that come in well-defined cycles—thus, many organisms have strong "circadian rhythms" that allow them to respond to light/dark cycles, and many coastal marine organisms move in anticipation of tidal rhythms. They may also use their keen sensory abilities and stored sensory observations to act in anticipation of unusual events, as evidenced by the animals responding to the December 2004 tsunami. This is not predicting an unknown future event but rather using sharply honed observational skills to respond to a challenge. Making and responding to predictions that are very unlikely to be correct is a waste of resources that are better spent finding food, avoiding predators, and mating. In the spring of 2011 alone, the devastating Japan tsunami with its commensurate effects on the country's nuclear power infrastructure, the so-called Arab Spring, and the outbreak of antibiotic-resistant *E. coli* in Europe were all threats to security that were possible to anticipate (along with an almost infinite number of other security threats that did not come to pass during that year) but impossible to predict.

Finally, a common misconception about evolution is that it is about seeking perfection, as encapsulated in the term *survival of the fittest*. This misinterpretation arises from misguided applications of Darwinian thought, such as eugenics, and it is reflected in more legitimate societal applications such as business performance analyses where "optimization" is seen as a laudable goal. In fact, evolution is neither about survival of the "fittest" nor about optimizing systems. How would you even begin to determine what is a perfectly adapted organism? While the Discovery Channel may spin the notion during "Shark Week" that the Great White Shark is the "perfect predator," wouldn't it be "more perfect" if it had deadly laser beam eyes? Indeed, trying to make a predictive science of optimality in organisms has proved itself to be almost comically wrong in case after case. For example, the first time depth sensors were attached to a live penguin, the animal dove to depths greatly exceeding its optimal performance as determined by mathematical theory and laboratory physiological experiments.[9] The successful results of evolution are organisms that are not perfect, but "good enough" to survive and reproduce themselves.

Eliminating the ability to plan, to try to predict, or to try to perfect would put a lot of military planners and business consultants out of business (it might also greatly reduce the number of PowerPoint presentations to which people are subjected). But organisms in nature have survived and thrived without these tools for billions of years because they have one powerful trait at their disposal—they are all adaptable. Adaptability is fundamentally different from merely *reacting* to a crisis (which happens too late) or attempting to *predict* the next crisis (which is almost certain to fail when complex ecological systems and human behaviors are involved). Adaptability controls the sweet spot between reaction and prediction, providing an inherent ability to respond efficiently to a wide range of potential challenges, not just to those that are known or anticipated.

This book is about learning how natural systems have used adaptability to survive in a risk-filled planet for billions of years and how we can harness this power ourselves right now. Using nature as a guide, we can observe how far our responses to security threats—whether they come from terrorists, viruses, economic catastrophes, or natural disasters—have missed the sweet spot of adaptability and how exactly we can get back there. Taken together, the suggestions in this book, developed from the perspectives of many different life scientists, soldiers, first responders, and activists, make up a system of "natural security" applicable to *any* situation where risk is present and unpredictable.

chapter one

THE ORIGINS OF NATURAL SECURITY

My transition from a tide-pool biologist practicing natural history to a security analyst practicing natural security is itself a story of adaptation. If you think about the first proto amphibians flopping out of a predator-filled sea to forge a new terrestrial life, it reminds us that adaptation requires leaving or being forced from your comfort zone and into a place where you observe and experience new threats to your security. In 2002, I left my comfort zone of marine biology research in the tide pools of the Pacific to work as a science advisor in Washington, D.C., for then Congresswoman Hilda Solis. There, less than a year after the 9/11 attacks, I had a window into the unfolding of a massive new security infrastructure. Although I was far from the tide pools where I had conducted my research, my inclinations as a natural historian led me to sketch the security environment of the nation's capital. At that time, it was impossible to be in D.C. without observing the sense of fear and desire for security that pervaded the Capitol Hill ecosystem. What I observed (to borrow Mike Davis's term for the environmental catastrophes of Los Angeles[1]) was an "ecology of fear." Jersey barriers continually emerged overnight like fungus in rings around monuments, museums, and government buildings. Mail arrived, uselessly, months after it was sent, brittle from the radiation treatments it had undergone in some Midwestern processing facility. Seasoned Capitol Hill staffers and young interns jumped tensely at any loud noise (which often turned out to be construction from the enormous bunker-like visitors' center that was being carved out under the Capitol building), and everyone kept portable chemical masks under their desks.

But something deeper troubled me about what I was seeing. While there was clearly *more* security in Washington, it was never *varied* security. It was as if all the new security measures in D.C. were installed as part of an animatronic diorama in the Smithsonian Museum of Natural History. Occasionally, new features were added—another fierce security guard here, another imposing concrete barrier there—and once in a while there was a change—security guards began systematically checking trunks of cars entering the Capitol parking lot after 9/11—but there was no variation in the elements of the diorama or their behaviors. The tide pools I studied in Monterey have inspired artists and authors and scientists not for their static perfection but for their continual movement and change, a variation that doesn't just provide beauty but is the driving force compelling the ecosystem and the creatures within it to continually adapt to keep secure.

As I walked through the routine Capitol security checkpoint each day, dutifully covering the keys in my pocket because everyone knew that's how you avoid setting off the metal detector, I kept asking myself, "How long would it take for a determined predator to get through these defenses?" Thinking about the practice of security (of which I knew almost nothing) in the context of biology led me to other questions almost every day in Washington. As I watched the debate over authorizing the Iraq war unfold on the House floor I asked, "Do animals declare war on other animals?" When the Total Information Awareness program was unveiled, with its spooky logo depicting the one-eyed pyramid from the dollar bill beaming its vision on the whole planet, I struggled to answer, "Which organisms in nature have total awareness of the threats in their environment?" As various versions of the Patriot Act were passed I wondered, "How does our own immune system profile invading organisms to determine who is benign and who is malicious?" As our defense and security budget grew ever larger, threatening to topple any semblance of a balanced budget, I asked, "How does a peacock survive when new predators colonize its home area?" And with each thwarted or successful terrorist attack I wondered, "Why does it make sense from an evolutionary standpoint to kill yourself in a suicide attack?"

As I asked these questions, trying to leap back and forth between biology and current events, I began to realize that not only did I know very little about security, I also knew very little about biology. In fact, what always appealed to me about a career in biology was the overwhelming diversity of things to study—I knew I'd never get bored—even as I knew any expertise I did gain would at best cover a tiny sliver of the whole picture. So I started asking my questions out loud, both to the security experts I met at Hill briefings and in committee rooms, and to my biologist friends around the world. Activating these two networks led to unexpected effects. Almost all of the security analysts and practitioners I talked to were thrilled about the idea of applying thinking and models from evolutionary science to security questions. They invariably confided in me that they were just running out of ideas much beyond dividing history into "pre- and post-9/11." By contrast, many of the biologists I talked to were initially skeptical that they had anything to contribute to a debate on security. Being an overwhelmingly liberal group, a few even flat-out refused to do anything that might legitimize what they saw as the Bush administration's exploitation of terrorism to justify its conservative agenda. But eventually, even some of the most skeptical came around, and almost everyone knew someone who probably had something to say on the matter, such that soon I had an informal group of ecologists, psychologists, anthropologists, paleobiologists, and virologists, not to mention security analysts, bio-warfare experts, and former (and, for all I know, present) spies, all energized to further explore the nexus between biological and societal survival.

I convened this group over the course of a year at the National Center for Ecological Analysis and Synthesis (NCEAS), a National Science Foundation–supported think tank in Santa Barbara, California, that typically supported meetings of scientists and economists to discuss big environmental science questions such as, "Why do certain species live in some places and not others?" "What is the global impact of commercial fishing on the world's oceans?" and "How much habitat must be protected to preserve species?" Fortunately, NCEAS stepped out of its comfort zone to support our working group, which I sowed with only a basic challenge: "What can we learn about security in society from security in nature?" Because of the enthusiasm of the participants and their commitment to step out of their own comfort zones, this initial question blossomed into an incredibly stimulating and creative discussion that continues to seed new lines of inquiry. As the ideas generated in several gatherings at NCEAS coalesced into key lessons, my colleagues and I have now begun to spread their conclusions to a wide range of audiences, from scientific societies and security think tanks to innovative companies such as Google and IBM, to elected officials and high-level security policy planners and boots-on-the-ground soldiers and first responders. Many of these audience members have become active participants in generating further refinements and applications of our initial explorations.

The ideas in this book thus come from many different sources of expertise and especially from the product of cross-breeding these lines of expertise. Yet the personalities of our working group members and the individuals we debated our ideas with were so strong and their ideas so compelling that it's impossible to let them fade into the interdisciplinary synthesis that resulted from our work. Accordingly, I will highlight several of their individual contributions in the chapters to follow. It is also intriguing to me that few of their ideas could really be separated in a coolly scientific way from their personal histories. Geerat Vermeij understands adaptive arms races in nature both because of his painstakingly detailed approach to natural history and because of the special adaptations his own brain and body have undertaken as a result of his early blindness. Luis Villarreal understands the deep evolutionary roots of self-identity systems (the mechanism by which all organisms know who is like themselves and who is different from themselves) both because of his expertise as a virologist and from his perspective as a mentor to young Latino science students. And Terry Taylor understands the value of symbiosis, not just because it's a good idea, but because his personal transformation from a soldier to a facilitator of peaceful solutions to conflict came through developing unexpected symbiotic partnerships.

But the primary driver of this book is what we can observe from nature itself. Observing nature is an endless task—one that can be tremendously enjoyable but also frustrating and confusing; just ask any birder. As with a field guide used by birders or other naturalists, it helps to have a few key things to look out for—identifying characteristics that help make sense of the vast diversity of nature. There are indeed a few simple themes that emerge in a study of natural evolution that are helpful to have in the back of your mind when considering different ways of applying lessons from nature to society.

FIELD CHARACTERS OF NATURAL SECURITY SYSTEMS

First (let's just get it out of the way), under the lens of natural history, humans are special, but not that special. There are a number of adaptations we have—such as advanced cognition and language—that both set us apart from most other species and create a lot of the complex security threats we face, but we are in the end just another species that evolved through time to deal with security challenges in our environment. With over a billion people facing chronic nutrition shortages[2] and a host of old and emerging diseases that threaten to turn into human pandemics, we are certainly still under pressures of natural selection. Moreover, the way we have evolved has changed our environment enough to force us to adapt further. This cuts several ways for us—we are extremely adaptable, but we also may have changed our world and way of living faster than some parts of us can evolve. Some of our adaptations, which first arose in a world completely unlike the societies we live in today, can get us into trouble now.

Second, just as humans are fundamentally similar to all other species, patterns in nature appear similar across different levels of biological organization. By levels of biological organization I mean the progression from molecules to DNA to cells to bodies of individual organisms to populations of those individuals to communities of those individuals interacting with individuals of other species to ecosystems that include the species, habitats, and chemical and energetic interactions between them all in a given area. What is remarkable is that similar patterns—for example, using noncentralized organization to sense and respond to the environment (discussed in Chapter 4)—appear at each level of this organization. Like Russian nesting dolls, biology has a *nested* quality. But biology is more than simply nested—each of the wooden dolls, after all, is just an independent entity, only connected to the others by having a similar shape and design. Biology is different because it is also a *recursive* process, meaning that the rules and patterns occurring at one level are not just similar to those at the next level but essential in defining what happens at the next level. In this sense, biology is like a spiral, and it shouldn't be surprising that spirals appear all over in nature—in the seeds of a sunflower, in the shell of a snail, and along the helical axis of DNA. All of this is a good sign for applying biologically inspired ideas to security in society because it suggests that solutions we devise that work at one level (say, within a single police precinct) will be applicable at a completely different level (e.g., throughout the Department of Homeland Security). It also invalidates the excuse that we can't change security policy unless our highest levels of government change. I will argue that we can start at any level of society in instituting more adaptable systems, and if we align our incentives correctly, these ideas can easily (in fact, will almost inevitably) spread up and down different levels of organization in society.

Third, complex natural patterns and processes arise from very simple building blocks. The four basic molecules of DNA code for a vast diversity of organisms that live in completely different ways and deal effectively with vastly different challenges. Natural selection, which has molded millions and millions of species into their forms today, is an incredibly simple process requiring just three simple elements—variation between individuals, environmental conditions that favor (or select) certain variants over others, and a means to reproduce those variants that are better suited to the environment. At higher levels of biological organization the simple process of individual organisms trying to survive and reproduce ends up producing networked ecosystems that are unpredictable, complex, resilient, and beautiful. Accordingly, natural security isn't about rising to the complexity of the security threats we face by designing a hugely complex system with flow charts and acronyms and multivariate statistical outputs. It's about finding simple processes that impart our security systems with the adaptability necessary to deal with a wide range of threats.

Fourth, good ideas in evolution are often identified because they appear nearly exactly the same across many different organisms. Although the DNA codes for millions of different organisms, the basic structure of the molecule and the process by which it replicates itself is the same across much of the living world. Heat shock proteins, which go around the body repairing damaged proteins, are another example, being both present and nearly identical in almost all organisms on Earth. I was amazed when I did some laboratory work studying heat shock proteins in marine snails that I could use commercially available heat shock protein antibodies purified from goats and rabbits to link onto and identify my snails' heat shock proteins. Many of the biologically inspired ideas I'll illustrate in this book are not just stab-in-the-dark guesses that happened to work out well, but time-tested billion-year-old solutions that have worked out in the coldest, highest, darkest, hottest, most predator-full and water-starved places on Earth.

Fifth, good ideas in evolution are also often things that evolve independently multiple times. Eyes, for example, are a good solution for finding your way around in a complex world, but there isn't one common type of eye that evolved billions of years ago and that we all share. Unlike DNA, this solution came later in Earth's life history, and it arose independently several times in different types of organisms. Octopuses have incredible eyes that serve the same kinds of functions as our eyes, but they are unique to octopuses. This phenomenon, called *convergent evolution,* is evidence that evolution is not about taking one design and plopping it down all over, but about solving problems particular to a given organism in a given environment. Throughout this book, I will propose ideas for security that mimic natural solutions, but they may have also been explored by other individuals or organizations which didn't make any reference to nature at all. I consider these to be examples of convergent evolution—different people trying to solve the problem of how to ensure security in society and coming up with similar solutions. Accordingly, the promise of a biologically inspired approach to security is not that any one finding from nature's security systems will be a revelation, unheard of among security experts, but rather that biology provides a holistic framework for simultaneously addressing many different types of security problems.

Finally, and most important, change and variation rule everything in nature. As Charles Darwin mused during his long journey on the *Beagle:* "Where on the face of the earth can we find a spot, on which close investigation will not discover signs of that endless cycle of change, to which this earth has been, is, and will be subjected?"[3] Darwin was referring to geology, the task he was primarily assigned during his fateful journey, but variation and change were very much at the heart of his subsequent biological studies. He felt it was essential to understand even the most minute variations—such as the microscopic differences between anatomies of the many species of barnacles that he cataloged in an enormous two-volume treatment[4]—to understand that "mystery of mysteries" of where life comes from. Later biologists would also come to the irreducible conclusion that variation and change were elemental features of nature. Edward Ricketts, the mid-twentieth-century marine biologist and philosopher whose ideas are scattered throughout this book, felt that the variation among organisms was the most important and ineluctable force in the natural world, noting, "Those residua, those most minute differentials, the 0.001

percentages which suffice to maintain the races of sea animals, are seen finally to be the most important things in the world, not because of their sizes, but because they are everywhere. The differential is the true universal, the true catalyst, the cosmic solvent."[5]

Thus, variation catalyzes change, change creates uncertainty, and uncertainty creates insecurity. In a world like ours, no effective security solution can be deployed and not modified or changed with time, because everything around it will be changing. Adaptation to these changes isn't easy. But the alternative—stasis—isn't acceptable in a world full of risk, variability, and uncertainty.

This book takes its guidance from the living world—the only set of examples that has consistently shown an ability to adapt, even to the harshest conditions. The millions of biological organisms on Earth have been adapting to environmental change and catastrophe for 3.5 billion years, a knowledge base unmatched by any human civilization. But before I sound too much like a used species salesman, let me reiterate that it's not the perfection of this biological history that I admire, but how these species have achieved so much as fundamentally imperfect beings. Biological organisms are like the hillbilly armored vehicles that Specialist Wilson drove in Iraq—cobbled-together collections of adaptations that get replicated when they get the job done, and eliminated when they fail. Accordingly, this book is not about how to develop a perfect solution to a given security problem, but rather about how to develop a flexible system for solving problems, however and wherever they arise.

This system of natural security won't be reliant on some new technological advance, although it will utilize technology when needed. It won't be designed by top government officials, but people at every level of society, including heads of major security agencies, can play a role in making it succeed. And perhaps to the consternation of the Central Intelligence Agency, natural security plans by their nature cannot be classified in any way. Rather, they are laid out in the structure of fossil and living organisms, in fragments of DNA, and in the observable behaviors of the organisms themselves.

This book uses the building blocks of life to address questions we should be asking about any situation where risk is inevitable and unpredictable in our environment. The book asks: What if we took a whole new approach to risk? One that didn't try to solve security threats piecemeal or only after they turned catastrophic. One that didn't waste resources on fixed responses that are useless against changing, intensifying, or intelligent threats. One that didn't over-prioritize one problem and leave the others unattended. One that rejects the notion that only a few elite "experts" on a certain issue are qualified to analyze it and decide how the rest of us should respond.

One place to start with this approach is to not be intimidated by the complex tangle our security problems have become, but to look for their simple common roots. Indeed, most security problems, large and small, stem from the same basic problem: the world is full of risk that arises from, and is exacerbated by, variation and uncertainty. The threats to our security today—whether economic, environmental, or existential—are often global risks that take on a huge variety of forms. There is huge uncertainty as to exactly how and when they can harm us or what we can do about them. Acts of terrorism have targeted individuals and large groups; used sniper rifles, bombs, chemicals, and airplanes as weapons; have had political, religious, and personal motivations; and have been committed by men, women, and children from our own and foreign countries. Cyberattacks can occur at any moment and can arise from a single computer or via an autonomous network of computers. And emerging infectious diseases thrive on the inherently variable process of genome replication, which allows suddenly deadly mutations to arise with little warning.

Even our limited successes against specific examples of these threats only serve to reinforce how difficult it is to control risk, variation, and uncertainty in the world. The eradication of smallpox, which was well worth the investment of about $1 billion in today's dollars,[6] did not eliminate the risk from infectious disease generally. Painstaking intelligence leading to the identification of members of a terrorist cell and their tactics, although critical to reducing the threat of terrorism, doesn't necessarily tell us much about another cell which may be metastasizing in another, very different part of the world. And the successful evacuation of shorelines after an earthquake, wisely undertaken because we understand something about the relationship between earthquakes and tsunamis, tells us very little about when and where the next earthquake or tsunami will strike.

Indeed, we might solve our current security problems if it weren't for risk, variation, and uncertainty. If all terrorist cells or all infectious diseases had the same characteristics, the tactics we applied to one would be effective against all. If tropical storms weren't intensifying due to climate change, levee walls built to withstand the "100-year flood" (estimated decades ago) would suffice to keep us dry. If risk wasn't ubiquitous, a one-time, all-out effort to eliminate it when it arose would be an appropriate, even if expensive, use of resources.

A geologist friend of mine wears a T-shirt that reads, STOP PLATE TECTONICS, which is a tempting way to think about eliminating the ever-present risk of earthquakes. In reality the T-shirt makes a funny (albeit a bit nerdy) statement because the notion of trying to rally public support to stop a global natural phenomenon seems absurd. Yet often we frame our efforts at solving global security issues in terms of eliminating their risk. Cybersecurity has struggled for over forty years to create perfect systems based on the model of "perimeter defense" that tries to keep all risk outside the system, and the result has been a cyberspace that is progressively less secure.[7] We have declared wars on terrorism and drugs—and, in my own field of ecology, a war on invasive species[8]—and continue to pour resources into them despite neither evidence that they are effective[9] nor a plausible argument for how victory over their underlying risks could even be declared. For example, on the one hand the idea of declaring victory over invasive species—including everything from microbes to algae and mollusks that clog up harbors and waterways to newly imported predators for which local species have no defense—is absurd in a world where anyone (and the microbes, spores, seeds, pests, and pets they intentionally or inadvertently take with them) has the potential to get to almost any other spot on the globe within thirty-six hours[10]—and millions do every day. On the other hand, if risk, variation, and uncertainty are inevitable, how can we possibly deal with the catastrophic threats that arise from these factors?

Fortunately, we have at our disposal a vast storehouse of largely untapped knowledge that could guide us in this seemingly intractable quest. It is a massive set of proven solutions, and teachable failures, to the very same problem that unites all of the threats we face—that is, how to survive and thrive in a risky, variable, and uncertain world. Remarkably, this database is completely unclassified and free to use by anyone. The solutions I'm referring to are all contained in the massive diversity of life on Earth—millions of individual living and extinct species, and countless individuals within those species—which have been developing, testing, rejecting, and replicating methods to overcome the challenges of living on a continually changing planet. These organisms have been experiencing security challenges and developing solutions since long before the latest presidential administration or Congress has been working on its agenda, since long before 9/11 finally woke most of us to the new post–Cold War reality, since long before industrialization pushed our biogeochemical cycles into chaos, and since long before humans ever walked the Earth. Indeed, the 3.5-billion-year history of life imbues biological systems with more experience dealing with security problems than any other body of knowledge we possess.

But because we have incredibly limited communication with all but one species of these millions and millions of natural security experts, how can we tap their knowledge? In some cases, we will have just the raw data to observe and work with—the diverse ecosystems, organisms, cells, and molecules, living in exotic locations and our own backyards, inhabiting our skin, and invading our genome. Still more knowledge can be gleaned from ancient observations of nature taken since the earliest human societies, from painstaking natural history and evolutionary biology conducted over the 150 years since Darwin's revolutionary *On the Origin of Species,* and from the most cutting-edge biological research on protein folding, genome mechanics, and network analysis that have massaged these raw data into stories and models and theories about how biological organisms survive and thrive on a dangerous planet. And because we ourselves are biological creatures, our own species' evolution and the modern manifestations of that evolutionary process are not only fair game but perhaps the most important set of data to consider. This means that in addition to the ecologists, paleontologists, virologists, and evolutionary biologists who have something novel to contribute to our security debate, so too do the anthropologists, psychologists, soldiers, and first responders who have extensive behavioral observations of people and societies under the stress of insecurity in an uncertain environment.

The questions that can be answered through observation of nature are the same questions that befuddle us in considering security in society: How can limited resources be allocated effectively between essential but competing needs? How can self-identity that drives individuals to react violently to outsiders—which appears throughout evolutionary history, from the earliest life forms on Earth needing a mechanism to identify viral invaders to the most recent radical religious sects—develop, thrive, and be broken down? How can hostile attacks be avoided or warded off? How can mutually beneficial partnerships be forged between entities that have wildly different and sometimes competing needs?

When we start to ask these questions of nature, one thing emerges above all else. That is, despite the massive variation in exactly how biological organisms keep themselves secure, all of these security solutions follow from one very straightforward concept: adaptability. Adaptability is the property of being able to actively and passively alter structures, behaviors, and interactions with other organisms and the non-living world in response to selective pressures. Selective pressures could be any change in the environment—climate warming, predators, a disease, limitations on available living space, competition for mates, overpopulation, failure of a food source. Organisms adapt within their own lifetimes (animals of the far north grow white coats in snow-covered months and mottled coats in the summer) and across generations (the evolutionary transition from aquatic reptiles to air-breathing amphibians) as those better-adapted variants mate more frequently or produce stronger offspring. In the loosest terms, adaptability occurs across all levels of biology. Individuals rich in "good genes" survive better. Species with a lot of well-adapted individuals stay on Earth longer. Ecosystems don't adapt as a monolithic whole, but those rich in adaptable species show amazing resilience to ecological change.[11]

Adaptability sounds intuitively attractive to anyone who has witnessed catastrophic failure of our security systems. Not surprisingly, many post-mortems of security disasters such as 9/11, insurgencies in Iraq, and Hurricane Katrina wholeheartedly endorse the concept of adaptability.[12] When I meet with security practitioners, from federal air marshals to TSA agents, Coast Guard, fire, and police chiefs, and FBI staff, it's clear that they have all been ordered, or have given orders, to be more adaptable in their practice. But "Be Adaptable" has been thrown around like a marketing slogan with few specifics to back it up. The problem is that on the one hand, government agencies have little experience or knowledge about how to be adaptable; and, on the other hand, although individuals—especially first responders and ordinary civilians in a time of crisis[13]—have shown remarkable adaptability, they often lack the resources or the power to replicate their adaptive behaviors across much larger swathes of society. Thus, their actions are seen as somewhat anomalous examples of grace under pressure—at best, they get a nod on the evening news as "real life heroes" before the commercial break.

This book intends to draw out from natural organisms, the undisputed world's experts on the subject, exactly how they incorporate adaptability into their own survival and suggest how we can, with our clever brains, deliberately incorporate adaptability into our personal and societal security systems without having to wait for the long march of evolution to get us there.

The real benefit of a biological approach is that it is the only framework I know of that can be applied consistently from security analysis to planning to implementation, and across the broad spectrum of security concerns. Nonetheless, my own limitations and my publisher's ensure that I cannot be equally broad in my coverage of natural security. There are necessarily security issues that get more (terrorism, homeland security) or less (food security, cybersecurity) treatment in this book, but my failure to adequately cover a particular issue should not be seen as any evaluation I've made as to which are the most important security concerns. Because I believe the same problems of adaptability apply to *any* situation where there is unpredictable risk, I have tried to use the most salient and sometimes the most provocative examples to illustrate various points.

My approach will arise from several different perspectives through the course of the book. At times I will simply lay out an argument, backed by both time-tested and cutting-edge research from a wide range of fields in ecology, evolutionary biology, psychology, anthropology, economics, and computer science, to make the connection between adaptation and security. Other times, I will let nature speak for itself (often through octopuses!), using the remarkable range of security measures taken by biological organisms to expand our own narrow sense of what is possible. And (usually playing the role of the hapless foil, but sometimes to demonstrate just how clever we can be) I will put human behaviors and the behaviors of our institutions (which often seem to be their own beasts) under the spotlight to expose how terribly far we may have strayed from our roots as a highly adaptable species.

There is no one model for how I present these ideas. Sometimes an animal model is completely sufficient to demonstrate the right way to make an adaptable security system. Other times, more subtle explanations are necessary. Sometimes humans are truly extraordinary and deserving of their place in the spotlight. In some cases, the translation from nature to humans will be straightforward. For example, DNA works the same in octopuses as it does in us. Other times, I will be making analogies that require a considerable stretch of the imagination or may simply not work for you, and that's fine. My goal is not to say that we are just like everything else in nature and therefore let's start acting accordingly, but to suggest that by opening our minds a bit to processes and patterns in nature we might find new ways to tackle long-standing challenges.

I hope the structure of the book in small part resembles adaptable biological systems, by focusing first and foremost on the problem at hand and then bringing to bear whatever tools are available to solve it. In a larger sense we will likewise have to throw out all of our preconceived notions and projections and focus directly on the immediate problems we face if we are to have any chance of creating truly adaptable security systems.

Nonetheless, because a book, unlike an ecosystem, can be planned, there is a logical order to my presentation. The next chapter begins our investigation, appropriately enough, in a tide pool. Like my scientific hero, Ed "Doc" Ricketts, who helped his friend John Steinbeck see that the whole world was contained in the tide pools of Cannery Row, I find that all the major lessons for natural security are there—oozing, fighting, stalking, and spawning in the tide pools where I got my start as a scientist—so they provide a good overview for the specific lessons that follow. I then drill down into one of the fundamental forces of life and adaptation, which is learning from the environment. Here I particularly stress the oddly undervalued power of learning from success. Both organisms and organizations learn to adapt better when they are organized adaptably. Chapter 4 untangles this tongue twister by identifying a consistent pattern in nature—the rejection of centralized control in favor of multiple semi-independent agents that individually solve problems as they arise in the environment. In Chapter 5 I show how nature turns the concept of redundancy—a term that instantly makes us think of waste, bureaucracy, and mass layoffs—on its head, because the type of "creative redundancy" found everywhere in nature is actually a vital force for survival and adaptation. In the subsequent chapter, I remind us that all this fabulous adaptable ability has consequences—it leads inevitably to adaptation amongst one's enemies—so that the history of life is peppered with episodes of escalation toward ever more clever and deadly adaptations and counter-adaptations. This escalation would quickly get out of hand if it weren't for channels of communication that have opened among and between organisms. Chapter 7 is about how information is used in the natural world to create uncertainty and to mitigate uncertainty, to create conflict and to avoid it. In Chapter 8 I get reflective, considering our own human adaptations and their roles, both positive and negative, in the security threats we face. Here it's the human animal, in both its typical form as a hunter-gatherer, and in its bizarre new manifestation as a city dweller living in an evolutionary cocoon, that provides the lessons for our continued survival. I then turn to the most powerful and underappreciated lesson for survival, which is that no organism, human or otherwise, does it alone. Perhaps because it initially seems so facile to say that cooperation is essential, we tend to overlook the power of cooperative, or symbiotic, relationships, yet they are everywhere in nature and vastly underutilized in society. In Chapter 10, inspired by the old swamp-dwelling proto-environmentalist cartoon organism Pogo, I remind us that we are our own worst enemy, wasting vast resources to create artificial and wholly inadaptable security systems right on top of living security systems that nature provided to us absolutely free. I close the book by acknowledging that security issues are only one small target for applying lessons from natural adaptability. Dealing with the effects of climate change, steering a business through a volatile world of bulls and bears, even teaching science to a troupe of K–12 students, are all areas where a static mindset will lead to devastating consequences and understanding adaptability may be the only way to survive and thrive in the twenty-first century.

chapter two

TIDE-POOL SECURITY

FISH DON'T TRY to turn sharks into vegetarians. Living immersed in a world of constant risk forces the fish to develop multiple ways of living with risk, rather than try to eliminate it. The fish can dash away from the shark in a burst of speed, live in places sharks can't reach, use deceptive coloration to hide from the shark, form schools with other fish to confuse the shark, it can even form an alliance with the shark, and all of these things may help the fish solve the problem of how to avoid getting eaten by the shark. But none of these adaptations will help the fish solve the general problem of predation, and it doesn't need to. The fish doesn't have to be a perfect predator-avoidance machine. Like every single one of the countless organisms it shares a planet with, the fish just has to be good enough to survive and reproduce itself.

Like the environment of fish and sharks, the world we spend our daily lives in is also full of risk. Acts of terrorism that seem to come out of nowhere. Wars that have carried on too long and show little progress toward resolution. Intensifying natural disasters fueled by global changes in climate. A distribution of food that leaves billions undernourished[1] and millions of others facing an obesity epidemic. A cyber infrastructure that we've become increasingly dependent upon that has also become increasingly vulnerable to catastrophic attack. New diseases and new mutations of old diseases that threaten to become global pandemics. The major threats society faces today are ominous and complex interplays of human behavior and environmental change, global politics and local acts of cruelty or carelessness, historical accidents, and long-simmering tensions. Some of these threats have plagued us as long as we have been human, and yet we've still made little progress against them; others are becoming more dangerous in synergy with rapid climatic and political changes; and still others are just now emerging.

Yet the responses we have been offered or forced to accept by the experts we've entrusted to solve these problems often seem frustratingly ineffective, naïve, or just plain ridiculous. When increased body screening of airline passengers was implemented after 9/11, Richard Reid attempted to destroy an airliner with a bomb in his shoe; and when shoes began to be screened in response to Reid's attack, al-Qaeda plotted to use a liquid explosive attack; and when liquids were banned, Umar Abdulmutallab used a powdered incendiary hidden in his underwear in an attempted attack. Far from any airport, on a tiny island in the tiny town of Beaufort, North Carolina, there is a tiny outpost of the National Oceanic and Atmospheric Administration (NOAA). Although I lack high-level security clearances, I'm fairly certain this little laboratory—which studies fish and coastal ecology—is not on any terrorist group's list of targets. Yet when the NOAA coastal scientists wanted to renovate and add some space a few years back, they were forced by the Department of Homeland Security to install enormous Walmart-style parking lot lights on their facility as a required security measure. This was ironic, since the scientists working at the lab know full well that nighttime light pollution is a major threat to the ecology of the same coastal marine environments that they are paid by taxpayers to study.[2]

Life on Earth has a lot to show us about how to create more adaptable systems than these, but with the doors to this vast pool of expertise on adaptability blown open, a daunting new question emerges: Where to begin? There are millions of species on Earth, each with its own special adaptations that themselves change within an individual's lifetime and over generations. Moreover, each individual of each species has its own unique quirks, aspects that make it slightly more or slightly less well-adapted to the particular environment in which it finds itself. In nature there are wings and eyes and claws and stingers and killer viruses and helpful bacteria, thirteen-year-long naps and 5,000-foot deep sea dives, suits of armor and solar energy factories, stolen poisons, and secret coded messages all working in some way to aid the adaptability and security of their owners. So, where would be a good place to start discovering which of these things would be useful to help solve our own security problems?

If you want to learn about security and adaptability from nature, I can think of no better place to start than a tide pool. It is the environment in which my career as a biologist developed, as a child during endless hours in sand flats and marshes of Cape Cod Bay, and later as a student and researcher in the great tide pools of Monterey and Pacific Grove. And it was in those same Pacific tide pools—where I retreated with my wife and infant daughter to get away from the horrible televised images on the morning of September 11, 2001—that I had the first inklings of an idea that they would be useful in studying security.

Tide pools are a good place to start studying natural security precisely because they are so full of life. The rocky shores of the North and South American Pacific coasts and the kelp forests that thrive just offshore are some of the most diverse and biologically productive ecosystems on Earth. And because they are so full of life, they are accordingly full of struggle for survival—millions of individual organisms each trying to compete for mates, find a preciously limited space to live in, forage for food, and not get eaten while coping with a dynamic and harsh environment.

Nestled between the edge of a continent and the fierce and relentless ocean, tide-pool organisms experience some of the most dangerous physical forces on Earth, and the greatest variation in those forces. At times the roiling surf crashes against the rocks with enough power to kill a person. A limpet, which is a small snail with a cone-shaped shell that lives abundantly on rocky shores, must withstand both the crushing pressure of these monster waves and the hydrodynamic lift that threatens to yank the little snail off the rocks as the water rushes over its body. Fronds of algae must grip iron-tight to bare rock yet be flexible enough to whip back and forth in the frothing brine, which brings them essential nutrients. A starfish uses thousands of suction-cup–like hydraulic feet to crawl around, pry open mussel shells, and devour them without getting washed to sea. The mussels themselves cling tenaciously to the rock by means of a few slender byssal threads, a remarkable natural material that makes human engineers crazed with envy because it is made in the relatively cold temperature of seawater and is at the same time stiffly resistant to abrasion and highly stretchable to accommodate a huge range of forces.[3]

Only a few hours later, when the tide recedes, the scene of that violent struggle for survival against physical and biological threats has transformed into a silent and serene little pool of ocean water trapped between the rocks. But as deadly as the high tide's waves might have been, the dryness, exposure, and searing heat of low tide may be worse. Now, on the high dry rocks above the isolated tide pool, many animals are hunkered down with a tiny store of water under shell or carapace to keep them alive. Seagulls and oystercatchers—and at night, the occasional cat, raccoon, or rat—are on the prowl, so the exposed animals need to hold on tight, hide in deep crevices, or settle into their home spot, a perfectly shaped scar they've scraped away into the rock with no room for a prying beak or scratching claw. Barnacles, which waved hand-like appendages to catch food in the ocean currents when the tide was up, now retreat into their citadel homes and slam four perfectly fitted doors shut against the sun and the predators. As the dryness and heat increase, all these organisms are responding not just with their bodies but within their cells as well. The proteins that allow them to carry out their daily activities are starting to melt, losing the precise structure they need to bind other chemicals and create metabolic reactions in the body. So while the organisms appear lifeless, they are frantically producing special "heat-shock proteins" that run around the body squeezing those failing proteins back into the right shape,[4] so they'll be ready to react when the cool water returns.

Below the surface of the tide pool, which at first appears as quiet as the rocks surrounding it, life is abuzz. I tell students on their first tide-pool visits not to move too much, but rather sit and watch the tide pool come to life. Soon something that looked like a pebble darts across the pool, a well-camouflaged tide-pool sculpin patrolling for a meal. Another quick flash reveals a shrimp, which doesn't need camouflage because its transparent body is nearly invisible in seawater. After some time watching, the comical circling and climbing and scuffling of hermit crabs resolves into something a little more sensible: the crabs, which protect themselves with borrowed snail shells, are sizing up any shell they can find, looking for a larger upgrade for their growing bodies. A flower-like anemone, lime green and fuchsia, waits quietly for its next meal to drift, or stumble carelessly, into its deadly tentacles with their thousands of harpoon-like stinging cells poised like an army in ambush. A glorious nudibranch—a snail that lacks the protection of an outer shell—ostentatiously winds its way across the anemone clad in its own stunning array of colors, and coolly stops for a meal. It will ingest some of those stinging cells intact, conscripting the little soldiers to use in its own defenses. The naïve predator that tries to eat the vulnerable-looking nudibranch will meet the strike of a stolen anemone stinger.

If it's an especially good day of tide-pool watching, you might catch a glimpse of the ultimate tide-pool survivor: the intelligent, secretive, deadly, and adaptable octopus. In John Steinbeck's famous *Cannery Row* (a novel that has been likened to a sociological study of a tide pool[5]), he describes the octopus thusly:

> Then the creeping murderer, the octopus, steals out, slowly, softly, moving like a gray mist, pretending now to be a bit of weed, now a rock, now a lump of decaying meat while its evil goat eyes watch coldly. It oozes and flows toward a feeding crab, and as it comes close its yellow eyes burn and its body turns rosy with the pulsing color of anticipation and rage. Then suddenly it runs lightly on the tips of its arms, as ferociously as a charging cat. It leaps savagely on the crab, there is a puff of black fluid, and the struggling mass is obscured in the sepia cloud while the octopus murders the crab.

Steinbeck's portrayal may seem awash in literary anthropomorphism, but modern video equipment has implicated the octopus in a real murder mystery at least as dramatic. Caretakers at the Seattle Aquarium were alarmed to find several of their sharks dead or missing week after week. An aquarist then stayed up all night with a video camera to reveal the culprit—an octopus that had been transferred to the shark tank (for which the caretakers were somewhat concerned for its own safety) had been leaping out at night to wrestle and squeeze sharks to death with its eight powerful arms.[6]

Octopuses learn not only how to survive, but thrive, in almost any environment. Even in the barren isolated tanks of a marine biology lab, colleagues have discovered octopuses escaping from their chambers and braving the dry air to scamper across a lab bench and find a snack in a nearby tank before returning to their own.

But this portrayal reveals only part of the octopus's success. With its soft meaty body, the octopus is an attractive target for predators. So it constructs a protective den in the rocks, sometimes with a peephole for its keen eyes to peer out from. If good rocky crevices aren't available, it will learn to use whatever is around it—a shell, an old crate, or the champagne bottle tossed decades ago from my advisor's shipboard wedding just offshore from the Hopkins Marine Laboratory in Pacific Grove. An amusing video making the rounds on the Internet shows octopuses in Indonesia that have learned to forage the increased numbers of coconut shells discarded from tourist boats and pull together two halves to make a spherical suit of armor.[7]

When the octopus does venture out from its constructed bunker, millions of cells on the surface of its skin are all sensing and responding to the world around, instantly changing shape and color to perfectly match their immediate surroundings. Once, after staring at a tide pool in Baja California for a long time, I thought I spied an octopus, but the small waves cresting the tide-pool walls riffled the surface too much to be sure. My eyes failing me, I reached my hand in to engage my tactile senses, and instantly a dark cloud of smoky ink filled the pool. By the time it cleared, I had confirmed my identification, but the beast was long gone. Steinbeck's good friend, the marine biologist Edward F. Ricketts (who is fictionally portrayed as the character "Doc" in *Cannery Row* and other Steinbeck novels), in his guide to the marine animals of the Pacific Coast said this of the octopus's capacity to blend in and hide: "A little observation will convince one that in a given area probably half of the specimens escape notice despite the most careful searching—a highly desirable situation from the points of view of the conservationist . . . and the octopus." He added, "The octopus has an ink sac, opening near the anus, from which it can discharge a dense, sepia-colored fluid, creating a 'smoke screen' that should be the envy of the navy."[8]

And if those defenses don't work, the octopus has a powerful jaw and a mean bite, as the ever curious Ricketts related in his field notes from a collecting trip in Canada: "Another big octopus. I've often wondered if octopi ever bite. Today I found out. Yes, they do, they certainly do."[9]

The blue-ringed octopus takes this biting to extremes. When threatened, it flashes dozens of brilliant blue rings across its body, a warning to predators that it is armed, in this case with highly toxic bacteria, powerful enough to kill a human, that live symbiotically in the octopus's salivary glands.[10]

But blending away like a wallflower at a dance, disappearing in a flash, and biting (usually) is no way to find a mate, and mating is as essential to survival as avoiding predators. For that, an opposite tack is needed—one that sets an individual apart from all others. Some octopuses, when spying a potential reproductive partner, will split their missions—the half of their body facing the mate will pulse with a psychedelic display of color, but the half facing the rest of the world (including other competing male octopuses) is dull and inconspicuous, as if to say, "Nothing special going on here."

Taken together, the octopus reveals almost all of the characteristics you would want in a biologically inspired adaptable security system. Its use of tools (the coconut shells) for future use and its well-known ability to wreak havoc on laboratory containment systems show that it can *learn from a changing environment.* The rapidly changing skin cells show it has an *adaptable organization* in which a lot of power to detect and directly respond to changes in the environment is given to multiple agents that don't have to do a lot of reporting and order-taking from a central brain. That it has an ink cloud *and* camouflage *and* a powerful bite that it uses both for offense and defense reveals its *redundant and multi-functioning* security measures. Its ability to deliberately stalk, surprise, and kill even prey much larger than itself shows that it can *manipulate uncertainty* for its own ends. Finally, its use of deadly bacteria in its own defense reveals that it *uses symbiotic relationships to extend its own adaptive capabilities.* Not all organisms in nature display these characteristics so prominently as the octopus, but all organisms use them to varying extents to survive and adapt.

APPLYING BIOLOGICAL IDEAS TO SOCIETAL NEEDS

If these characteristics of adaptability have served millions of organisms well for billions of years, they might be worth learning from, given that we've only inhabited this Earth for the tiniest fraction of that time. Yet throughout history the translation of biological ideas to societal affairs has been met with distrust or resistance. Ed Ricketts himself made several forays into this arena. In fact, Ricketts, with his boundless curiosity (not to mention a steady supply of Prohibition-skirting alcohol ostensibly used for pickling specimens, but occasionally redistilled for libations), attracted a remarkable group of early-twentieth-century writers, poets, philosophers, scientists, and musicians to his tiny Cannery Row laboratory, all interested in exploring the connections between the forces of nature and society. For John Steinbeck, who spent long hours in the tide pools of California and Mexico with his dear friend Ricketts, there was an inextricable link between the struggle for survival in the tide pool and the struggles of characters like Tom Joad and his family in *The Grapes of Wrath,* fleeing the environmental catastrophe of the Dust Bowl and straight into a fierce social struggle in the agricultural fields of California. For the mythologist Joseph Campbell, an early collecting trip with Ricketts to the Pacific Northwest, which included encounters with native Canadians fighting to hold on to their traditions amid rapid societal change, inspired and guided his quest to find universal characteristics of creation and hero myths within human belief systems.[11]

Ricketts's own ideas connecting nature and society were both pragmatic and transcendental. He saw the immediate value of nature study to the emerging war effort. He noted wryly in his journals that the Japanese fishermen he had encountered on his collecting trips along the Pacific coast of North America in the late 1930s had perhaps been taking a few more depth soundings than were necessary for shrimp trawling.[12] At the outset of World War II he offered to the U.S. Navy his understanding of the coral reef biology and oceanography of the Mandated Islands of the Pacific, which he felt would be invaluable for any marine invasion there. But the navy never responded, and he gradually accumulated evidence that the military wasn't ready to use much natural history data in its war effort. He noted in his journals the vain efforts of a colleague to help the navy understand fouling on their ship hulls by barnacles (which drift in the water as microscopic plankton until they find a hard surface to settle upon and transform into the shelled adult forms that greatly reduce the speed and maneuverability of ships):

> At the start of the war, she heard the navy had a ship fouling project near Bremerton. Offered her services. Chiefly patriotic.

> But the navy was very polite and stiff-necked and stuffy. Pure scientists, biologists especially, no doubt had their uses, but the navy wanted practical people. Engineers certainly, but no theoreticians. Biologists, especially systematists, were ivory tower. A year later the navy's long and costly experiments had been ruined. Curiously there was no difference between the expensively set-up experiments and the controls. In fact the navy had been so perfectionist and so practical that they had filtered every bit of water used in the experiment. At the same time filtering out the larvae that even the most abstruse biologist could have told them would have developed into the fouling organisms they wanted to deal with![13]

In this case, it was the navy's search for perfection that did in its experiments—its engineers filtered out all the variation and messiness of life itself, and in the process lost track of exactly what they were looking for. While the example appears like a caricature of itself, it sadly reflects our continued search for perfect security solutions, rather than ones that solve the problems at hand.

But Ricketts's efforts to bring biological understanding to human conflict stretched far beyond nautical charts and biofouling. He struggled his whole life to outline a philosophy of "breaking through," through which he sought to define the conditions under which people in conflict might transcend their differences and find a new, mutually beneficial state of understanding. Central to this philosophy was the idea that people could break through conflicts if they set aside their preconceived notions and biases about the other side and came to a dispute "honestly presenting the main theses, however controversial."[14] This followed directly from his study of tide-pool organisms, for he felt that the only real difference between the struggles of society and the struggles of tide-pool organisms was that the tide-pool creatures more straightforwardly revealed their intentions through their actions:

> Note the relations between individuals, families, friends, races, and nations, which involve expediences both political (power-driven), emotional and economic (hunger and poverty), with their ideologies, competitions, needs, overproductions, overpopulations and wars. Most of these have their primitive counter-parts along the shore. Who would see a replica of man's social structure has only to examine the abundant and various life of the tidepools, where miniature communal societies wage dubious battle against equally potent societies in which the individual is paramount, with trends shifting, maturing, or dying out, with all the living organisms balanced against the limitations of the dead kingdom of rocks and currents and temperatures and dissolved gases. A study of animal communities has this advantage: they are merely what they are, for anyone to see who will and can look clearly; they cannot complicate the picture by worded idealisms, by saying one thing and being another; here the struggle is unmasked and the beauty is unmasked.[15]

Ricketts himself was heavily influenced by his only academic mentor, the early ecologist Warder Allee, who Ricketts knew through his short and incomplete undergraduate training at the University of Chicago. Allee was particularly interested in the cooperative relationships between organisms, and he saw them as a model for human society. At the outbreak of World War II, Allee also wanted to contribute his biological knowledge to the security challenges of his day, lamenting, "The present system of international relations is biologically unsound."[16]

But if these efforts to connect biology and society were ignored as academic musings before and during the war, the postwar twentieth century saw them actively dismissed as unscientific or even dangerous. In part this was due to a perceived relationship between eugenic philosophy practiced by the Nazis and interpretations of Social Darwinism centered on the concept of "survival of the fittest" (to be clear, evolution is not about survival of the "fittest" at all, but about survival of the "just good enough" to reproduce). At the same time, the life sciences were going through an absolute revolution, brought about by the discovery of the structure of DNA. The old plodding way of studying biology by building intensive repeated observations of nature into grand sweeping syntheses was pushed into the back corners of academia, and even hugely accomplished biologists such as E. O. Wilson at Harvard found themselves marginalized by the giants of the new molecular biology.[17] In response, the study of ecology sought to be more like its successful molecular cousin, looking for clean and absolute laws hidden in mathematical approximations of nature and designing small-scale experiments that provided clear and unequivocal evidence of some ecological phenomena, no matter how trivial. The notion of just going out and observing nature for its own sake was relegated to the dalliances of hobbyists. High schools, colleges, and even graduate programs stopped teaching basic zoology, botany, and taxonomy,[18] and funding for long-term programs to simply monitor changes in species populations dried up.

By the end of the twentieth century, the idiocy of turning our backs to nature became unequivocally clear. The erosion of natural history education occurred alongside an alarming trend toward children spending continually less time outdoors, which may relate to the higher incidences of childhood obesity, inability to focus, and mental health issues—a suite of problems Richard Louv has termed "Nature Deficit Disorder."[19] More broadly, the organisms studied by ecologists in tiny experimental treatments or in the laboratory were disappearing throughout their natural ranges, whole bodies of water were becoming "dead zones," even the entire climate was being changed, and that in turn was completely reorganizing ecosystems.

Because these complex issues inevitably involve human behavior—burning fossil fuels in our cars, letting excess fertilizer wash down the Mississippi, fishing a species until not enough are left to make fishing economically viable—the recursive path tracing the relationship between the activities of organisms in a tide pool or a savannah and the activities of humans in societies is more clear than ever before. Accordingly, the application of biological ideas and evolutionary theory to societal questions is once again becoming accepted by scientists[20] and welcomed in society for the novel insights that often come with it. Paleobiologist Geerat Vermeij, a MacArthur Genius Grant recipient who has been hugely influential on the work discussed in this book, has recently demonstrated that the economy of society and the economy of nature, including concepts of trade, competition, power, and collapse, are built on the same principles.[21] Applying this analogy in the wake of the recent financial crisis, both the New York Federal Reserve Bank and the Bank of England have consulted with biologists to glean ideas from biology, evolution, and ecology on how to manage the complex global "ecosystem" of finance.[22] "Darwinian medicine," which considers the evolutionary roots and causes of disease and shifts treatment strategies accordingly (for example, by recognizing the adaptive value of most fevers in fighting infection rather than seeking to suppress them with drugs), is slowly taking root.[23] A whole field of "biomimicry" has emerged, which collectively looks at how computing, energy generation, architecture, industrial agriculture, and medicine, among other fields, can be dramatically improved by copying and adopting nature's strategies in these areas.[24]

The old traditional approaches of social-biological study—that of Allee and Ricketts and Steinbeck—in which careful observation of nature provides the essential building blocks of understanding larger, more holistic societal issues, are more valid and essential than ever. But the translation between natural systems and human society cannot be, and should not be, a perfect cloning. As a scientist who has worked in Congress, I have learned that legitimate ethical concerns, shady politics, and everything in between make the literal translation of many scientific ideas into policy both impossible and, usually, inadvisable. Nonetheless, we have erred too far in the direction of ignoring biology altogether, leaving abundant room to incorporate biological wisdom into whatever logic currently drives our security analysis, planning, and practice.

Making this science-to-society transition work will require a mutual learning process by anyone who wants to get past the "worded idealisms." And, as I show in the next chapter, a learning process is really the start of an adaptable process. Learning is essentially a force of nature that makes adaptation possible. For organisms in nature or organizations in a bureaucracy, harnessing the power of learning from environmental changes is a key first step in becoming more adaptable to changes that will inevitably come in the future.

chapter three

LEARNING AS A FORCE OF NATURE

LEARNING IS AT THE CORE of all security situations and responses. Nearly every security situation arises because of some past failure to learn by one of the parties involved, and nearly every security situation requires learning to respond to the situation and to prevent it from happening again. Learning about security in nature takes a wide range of forms. Learning can involve a single dramatic lesson that becomes entrenched for life, or it can be the outcome of an ongoing process of selection and adaptation.

When my family and I moved to Tucson, Arizona, we took our dogs, who were inclined to chase anything that moves, to "rattlesnake aversion" training. They were brought right up to a live but defanged rattler, and when they got within striking distance, the trainer applied a mild—but surprising—electric shock. The dogs howled and leapt back in terror from the snake. Then they were shown another, larger snake in a cage and given a second shock. After that, just seeing a snake without any shock sent them jumping away or cowering behind my legs. The trainer, who gets a nifty $65 per dog for the five minutes the training takes, guarantees this associative learning will be with them for life.

The problem with such associative learning is that it only deals with a specific threat. After the rattlesnake training, our daughters, who were themselves learning about the dangers of desert living, asked, "What about scorpions? Do we have to take the dogs to scorpion training now, too?"

Much of our security systems seemed based on associative learning, and this leaves us vulnerable to an enemy that can learn. For example, after the anthrax attacks of 2001 we developed an association between U.S. mail and terrorism, and we responded not unlike my frightened dogs. Years later we still irradiate mail bound for the Capitol and greatly restrict uses of curbside mailboxes. Yet it's not hard to figure out how to deliver a letter bomb using the U.S. Postal Service. I learned last fall when I tried to send a package from Santa Barbara to Los Angeles, a distance of 96 miles. Three weeks after I dropped the package into a curbside mailbox, it appeared in my former office at Duke University in North Carolina, 2,613 miles away. A sticker affixed to it told me that the package had been returned to the sender because new security measures required any package over 16 ounces to be handed directly to a postal worker. For me, the sticker on the package represented a minor annoyance. To a terrorist, it would read like a schoolhouse primer for creating a disaster. Just put your target's address in the return address field and the post office will gladly deliver it. Best of all, if you are feeling really cheap, you don't even have to put postage on it.

A high level of punishment (an electric shock, a terrorist attack) or reward (hitting a jackpot on a slot machine) leads to quick retention of associative learning, but it's not necessarily adaptable learning. Learning to adapt is a much more complex process that takes place across a number of planes—individual or organizational, passive or active, through the ritualized transmission of knowledge or as an ongoing experience in a changing world. Under this broad view of learning it appears to be a *force* of nature, like gravity. That is, the transmission of information, and the resulting alteration of interaction with the environment by the receiver of that information, occurs in nature no matter what you do. It may be encoded in DNA, passed on from a mother sea otter to her pup, or written in an al-Qaeda field manual, but learning is at the heart of adaptation.

Viewing learning as a force of nature leads us along the same path that we take when we recognize risk as ubiquitous in nature—whether we chose to learn well or not, our enemies will learn, so we'd better learn well and understand how learning relates to adaptation. The postal security measures fail to account for the fact that learning in nature sets off a continual process of escalating threats and adaptive defenses.

Humans tend to get a bit provincial about the subject of learning. We tend to assume that beyond some silly experiments with mice in a maze, there isn't too much we can learn from the nonhuman world about learning because we are the undisputed world champions of learning. Moreover, one might argue that the vast majority of us, certainly those of us reading this book, do not live in the tooth-and-claw world of animal struggles for daily survival, so the kind of things that we assume animals need to learn are not that relevant to us. I would argue, first, that while we are remarkable learners, we are not that different from many animals in our relationship between learning and adapting to a changing world, and second that it is the long-term *process* of adaptable learning—conducted throughout a lifetime and across generations in all organisms—from which we can learn richly from nature.

Almost all modes of learning in humans have been identified in other biological organisms. Tool-making was once considered the province of humans alone, but then chimpanzees were discovered making specialized tools to fish out termites from rotting logs. Then those who would separate humans from other animals moved the goal line and suggested that deliberately making tools for future use (rather than just to extract a tasty treat or deal with a threat near at hand) was the barrier. But a chimp named Santino in Sweden's Furuvik Zoo methodically breaks apart the cement in his enclosure, calmly stacks the pieces in a special place, and only later, when the visitors arrive, does he use these tools by angrily hurling them at the surprised tourists.[1] Critics may take solace in the fact that this occurs in an animal that is extremely closely related to humans, but now we have film of octopuses, which are about as genetically close to us as a garden snail, collecting different halves of coconuts *when they are not threatened* to use in constructing a rolling suit of armor when they subsequently become threatened.

Another supposed barrier between human and non-human learning was confidently determined to be the sense of "self" supposedly unique to humans. But primates in captive learning environments have clearly shattered that barrier. And the Indonesian mimic octopus, which was only discovered in 1998, can quickly mimic, in both appearance and behavior, at least a dozen other animals like flatfish, sea snakes, and deadly lion fish.[2] Does its ability to change itself into something else entirely indicate it has a sense of itself as something that must be transformed to appear like those other selves? We'll probably never know for sure, but the more we look, the less substance we find to the categorical distinctions between humans and non-humans.

The adaptable learning *process* at its most fundamental level is driven by natural selection as envisioned by Darwin. The basic tenets of Darwinian evolution are threefold and quite simple. You need variation (for example, two birds are hatched to the same parents—one is solid yellow and one is yellow with mottled gray spots); you need a selective force that literally favors one type of variation (a predatory hawk that easily spots the yellow chick but not the mottled one); and that selection needs to punish the inferior variant (the solid yellow bird dies) or reward the superior variant (the mottled bird grows up and reproduces). This process is not the only way that new species arise and adapt,[3] but it is an extremely important part of the ultimate domination of a once-lifeless planet by a stunning diversity of life. This chapter is about how we use and misuse selective pressures to learn or not learn lessons from our changing security environment.

Learning is intricately related to adaptability in both the process of how it comes about and in the results of the learning process. One of the fundamental properties of selective learning in animals is learning to recognize signals from the environment such as the color of the sweetest flowers, the appearance of edible versus nonedible prey, or the songs of potential mates. This learning is selective because those organisms that simply can't identify what is good to eat, what is good to avoid, and who is good to mate with will die out or fail to pass on their genes.

But simply recognizing these signals is not enough, because it doesn't account for the ubiquitous variation and change in nature. An organism that can only identify a particular signal will starve or get eaten or fail to mate if the critical signals related to feeding, predation, and mating change. So organisms critically learn to distinguish small differences between signals—the sweeter sweetness of a deeper red flower, the most palatable prey among a herd, or the more robust call of a stronger male mate. And remarkably, it is from the development of this finely tuned discriminatory ability that animals learn the most important skill of all—the ability to generalize their sensory abilities and be able to respond to novel stimuli from environments they have not yet experienced.[4] In other words, through the process of learning in detail about everyday threats and opportunities, organisms gain an adaptive capability that serves them in responding to future unknown threats.

Even the process of how animals learn is not immutable. Darwin, who fought against his contemporaries' notion that animal species were immutable—unchanging since the dawn of creation—would not be surprised to read recent studies demonstrating that within lifetimes and across generations, animals can learn to learn differently. That is, animals have some basic capacity for learning, but they can learn in accelerated ways depending on the environment they are put in. Monkeys, which are generally considered to have the learning capacity of a human two-year-old, can be trained in experimental settings to learn like a nine-year-old, including understanding a sense of their own self as a unique entity interacting with and affecting the world around them.[5]

The animal world thus tells us that learning is itself a selective adaptation; that learning to discriminate variation in the immediate world helps to build an adaptive capacity to respond to novel threats; and that the process of learning can change, given the right conditions. This powerful suite of properties should be seen as a warning, reminding us that no security adaptation should be assumed to be a safe everlasting solution, because there is always the potential for an adaptable enemy to learn how to overcome it.

Nonetheless, the degree of difference between humans and others is sometimes impressive and bears special attention here because human learning plays such an important role in protecting, and threatening, our security. Humans possess a wide variety of learning abilities, and they learn fast, especially when working with or against other humans.

HOW HUMANS LEARN

Michael Kenney, who has extensively studied the ongoing battles between narco-traffickers and international law enforcement efforts in Colombia and has applied his knowledge to the battle between terrorists and counter-terrorists, believes that learning is the key mechanism for adaptability on both sides of these battles. The different ways in which learning occurs among individuals and organizations is the most telling indicator of how well they adapt to changing environments and to the activities of their adversaries.

Kenney uses the Greek terms *techne* and *mētis* to differentiate two different modes of learning. Techne refers to formalized knowledge—facts, figures, techniques, plans, and other data that can be conveyed through lectures, field manuals, and other academic training. Mētis, by contrast, refers to knowledge that is gained through experience and interaction, learning through failures and successes during day-to-day operations, and spread through networks of practice. Kenney finds that while both techne and mētis are essential forms of knowledge transfer, successful criminal or law enforcement agents primarily rely on mētis to maintain the capability to adapt to continually changing environments:

> Like drug trafficking, counter-drug law enforcement is a mētis-based activity that requires improvisation and responsiveness if it is to be successful. Effective agents often supplement their formal training with practical know-how and intuition that comes from performing their activities repeatedly in local environments. Like the narcos they seek to apprehend, law enforcement "narcs" cultivate mētis in carrying out investigations and adapting their activities in response to unanticipated circumstances.[6]

But where do these advanced forms of technical and experiential learning derive from? Are we born with them, or do we learn to learn? Alison Gopnik, a psychology professor at UC Berkeley who studies childhood cognitive development, argues that while both technical learning and complex experiential learning are made possible by the brain morphology and physiology that all humans are born with, they only attain their true potential through an adaptable learning process that begins in the rapidly changing brain of an infant. According to Gopnik, one of the signatures of human learning is the extreme capacity for change that accompanies every level of human development. In contrast to the popular belief that babies possess incomplete or quasi-defective adult minds, Gopnik argues that babies have a fully functioning, but radically different, mind than that of a toddler, a teenager, or an adult. This mind is protected—by a heavy degree of parental care—from having to do all the everyday things like hunting, gathering, and balancing checkbooks that adult minds do, and instead it is hard at work mapping how the world works and developing adaptable learning strategies that will allow it to respond to completely novel experiences it will ultimately face. Other species that have extended periods of parental care—crows, for example—also show adaptive capacities to learn, but no species spends as much time caring for its young—and thus allowing adaptable learning systems to develop—as humans.[7]

As babies grow up and catalog more experiences, the human mind increases its store of memories. But memories aren't just nostalgic photo albums of the mind. They help us quickly understand the present and even predict the future. Our memories are essentially chunked as packages of related observations. These chunks aren't held in a specific address in the brain but in a networked set of neurons that fire in a certain pattern when the memory is invoked.[8] These little programs form recognizable patterns that can be recalled even in completely new situations. A chess Grand Master has on the order of 50,000 positions he can instantly recall, and he can also have the ability to remember significant setups—the moves that deliver players to that particular position.[9] We now have computers, which through sheer computational muscle and through being dedicated solely to the process of playing chess, can competitively play against Grand Masters, but they don't play *like* Grand Masters.

A computer can grind out calculations, but it can't predict the future. However, experienced observers who have chunked their memories can essentially do this. Before he was famous for writing books like *The Tipping Point* and *Blink,* Malcolm Gladwell wrote an article for *The New Yorker* on "Physical Geniuses," people like the great hockey player Wayne Gretzky and the cellist Yo-Yo Ma.[10] What Gladwell found was that these people put in an enormous amount of time practicing their craft, and all that practice gave them a huge storehouse of chunked memories that could be recalled consciously or subconsciously and used, even to predict the future. When Gretzky played hockey, he had an uncanny ability to make the perfect pass to set up a goal (although Gretzky holds the all-time NHL records for both goals and assists, he made nearly twice as many assists as goals). Gladwell showed Gretzky a film of one of his perfect passes that set up a goal and asked The Great One to explain how he knew his teammate would be in the right place. Gretzky struggled to explain it, but basically acknowledged that given all the conditions and positions of the players on the ice right then, he knew his teammate would be where he was sending the puck. It is difficult to describe this kind of learning consciously with words, both because chunked memories live in the province of the subconscious—we use them before our conscious mind even knows we're using them—and because we generally don't feel too comfortable talking about an ability to predict the future.

ORGANIZATIONAL LEARNING

If individual humans can learn in complex ways, it would seem to follow that organizations created by and composed of individual humans would also have the capacity to learn. In nature, learning occurs at multiple levels of organization. An immune system within an individual organism learns each time it interacts with a pathogen. Individual organisms then learn through a combination of their inherited genetic code and changes they make throughout their lifetimes in response to changes in their environment. Their individual learning may then impart a higher-level learning as the species they belong to becomes enriched in or depleted of individuals that operate a certain way. This multi-level aspect of learning suggests a tight correspondence to individual and group learning in society. Just as species "learn" from the collective successes and failures of lower orders of biological organizations, it would seem that organizations should learn from the actions of their individual members.

Yet we often decry how societal organizations and bureaucracies don't seem to learn any lessons. When high gas prices followed by a collapsing economy led American car buyers to abandon the once popular and enormous gas-guzzling SUV trucks and giant automakers like GM were brought to the point of needing a massive government bailout, people wondered if they had learned anything from the gas crisis of the 1970s, when smaller fuel-efficient Japanese cars began to flood U.S. markets and gained a permanent stronghold. Organizational inertia in government seems to prevent us from responding even to known threats, such as terrorists' clear intentions to do harm to American targets, until a catastrophe occurs.[11] But there are also many examples of nimble organizations that appear to learn from their own past as well as from the experiences of other organizations, and there is a strong interest in understanding how they do it. What makes an organization learn or not learn is the stuff of CEO and shareholders' dreams and nightmares, and accordingly, a vast popular and academic literature has arisen around the topic of how and if organizations learn.

At first blush, much of the organizational learning literature seems to take pieces from a biological playbook, and sometimes it comes tantalizingly close to identifying a kind of organizational learning that is essentially biological. Most of the papers on organizational learning that I've come across take some riff on the *variation* theme, noting that the business environment is constantly changing and businesses must change their practices accordingly. This literature also generally acknowledges that some force of *selection*—typically in the form of enlightened managers (presumably those who suffered through that particular article or organizational learning seminar) who must screen employees and departments to make sure that they are learning—is necessary to separate out learning organizations from those that don't learn. And a subset of the literature also pays deference to *replication* of successful variants—noting that rewards to departments and individuals for learning effectively are necessary to keep an organization learning in the future.

Yet the literature quickly strays from the organic world into theoretical constructions and flow charts full of various boxes representing entities like "organizational routines" linked by arrows representing actions like "organizational double loop learning"[12] that supposedly map out how organizations should optimally learn.

Even the "realists" in the business literature, who scornfully reject the boxes and arrows in favor of collecting real data about real companies, remain too tightly constrained by their business backgrounds to capture the full extent to which learning is linked to adaptation, as it is in nature. In part this stems from a misappropriation of Darwinian thought that is still omnipresent in business thinking—the notion that success comes through "survival of the fittest." This kind of thinking leads inevitably to the idea that a successful organization must *optimize* every component of its practice as it strives for perfection. In turn, this pathway requires constant benchmarking and comparison to a set of predetermined metrics.

Harvard Business School professor David Garvin, who was among the first to call for more realism in the organizational management discussion, backed up his call with a detailed quantitative benchmarking tool that any organization could use to measure just exactly how prone to learning each component of its organization is.[13] The model asks employees dozens of questions in each of several "blocks" related to the organization's learning environment, learning-related processes, and leadership, and scores for each sector are measured on a 100-point scale. The organization can then compare its scores to benchmark averages of previously tested organizations and presumably work to improve the learning capabilities of each of its underperforming sectors.

The problem with this kind of approach is that it assumes there is a value of learning for its own sake. By the time you get to the level of measuring each of what a researcher has determined to be the most important individual components of learning, you may have lost sight of why it is important to learn in the first place. That is, having the capacity to learn does not mean an organization will actually learn or learn the right lessons, any more than having a Ferrari in your garage makes you a great driver.

Robert Wears, an emergency medicine researcher at the University of Florida, noted this when commenting on the organizational learning literature. He cites an example of a medical article that lamented the lack of hard evidence for learning among clinicians treating shock in children, even as the article reported a tenfold *decline* in children's shock deaths over the same time period![14] At a conference on adaptability in military operations that I attended, a senior naval analyst pleaded for new ideas on how to better train troops to learn to be adaptable despite being unable to identify where or when troops on the ground were failing to adapt. Natural organisms don't need to "benchmark" learning because nature makes it abundantly clear when learning is needed. Likewise, extreme circumstances in human societies, such as are found in emergency medicine and warfare, demonstrate their own unequivocal results on how well individuals and organizations learn.

A further assumption in trying to benchmark learning is that there is actually some desirability in optimizing each component of learning. Nature doesn't give a fig for survival of the fittest, nor for optimization. If an organism is surviving fine and reproducing its genes through the range of environmental variation it experiences, then good enough. Although it is taken to have an objective, technical connotation, "optimization" in the business literature is a value-laden term, like "ruthless" or "cooperative" or "creative," that appeals broadly across a wide spectrum but doesn't necessarily indicate a useful response to a difficult environmental problem.

In natural adaptive systems, value-laden terms need not apply for acceptance—an organism might be successful by cooperating, by altruistically helping its kin, or by building structures that other organisms rely on, or it might eat its own sisters, serially rape females, and dump toxic chemicals all around it, making the environment entirely inhospitable for everything else nearby—it doesn't matter as long as it passes on its genes or helps a closely related individual pass on its closely related genes. Here I want to reassure you that it is *not* this value-less character of nature that I am advocating we replicate in society—I strongly believe that all sorts of values—ethical, economic, political, and social—necessarily come into play when changing policy and practice. Moreover, to be more precise, when it comes to organisms such as humans (and likely a number of other animals with advanced cognitive capacities), values *do* interact with the day-to-day struggle to survive, so that the net effect is that values can play a role in some evolutionary systems. Rather, I am suggesting that the truly important characteristics of natural learning systems—their ability to integrate all sorts of experiences, past, present, and hypothetical future into patterns—can be inculcated in society without resorting to value-laden goals disguised as objective benchmarks.

It just doesn't matter how close the organism is to its own theoretically optimal performance. It might work at 25 percent of its capacity and still survive just fine in a given environment. An old acquaintance with a business background and an indefatigable entrepreneurial spirit once made a small fortune selling enamel lapel pins that stated simply, "110%" as "motivational gifts" for businesses to give to their employees. I doubt he would have done so well with pins that read "25%" or "Just Do Good Enough." Yet in nature, the organism that gives 110 percent or even something close to 100 percent of its capacity to a given task is almost assuredly going to wind up dead.

FORCES OF LEARNING

Natural learning is neither a deliberate nor an optimized process. Indeed, the vast majority of organisms—even humans most of the time—learn without being consciously aware that they are learning. Learning carries much of the same qualities of Darwinian evolution—it is made possible by selection for, and reproduction of, those variants who learn the right lessons from the environmental challenges put before them.

These selective forces are likely at work behind a disturbing global trend that has been acutely experienced by U.S. and allied forces in Afghanistan and Iraq—namely that the objectively weaker sides of conflicts, in terms of technology, firepower, troop numbers, and financial resources, have become over the past 100 years or so increasingly likely to win wars.[15] There are a number of reasons why this may be the case. In Vietnam, for example, both the greater familiarity with the terrain and acclimation to the climate were certainly a great help to the Vietcong. But evolutionary biologist and security analyst Dominic Johnson points to Darwinian selective forces as a likely culprit behind the success of weaker sides. The reasons for this lie in all three components of Darwinian evolution: variation, selection, and replication. First, insurgencies fighting regular armies tend to have a more diverse set of tactics at their disposal, whereas the regular armies they fight are constrained by long-standing institutional norms, ethical and legal constraints (such as the Geneva Conventions), and standard operating procedures (that themselves have become such an integral part of military operations that they are routinely referred to by their acronym: SOPs). In Iraq in particular, insurgents were also drawn from a much more diverse population than U.S. forces: the 311 foreign fighters captured in Iraq between April and October 2005 came from 27 different countries.[16] The differences between an army specialist from Poughkeepsie, New York, and one from Lubbock, Texas, pale when compared to the variation between a hardened fighter from Sudan and an eager new al-Qaeda recruit from Syria. Second, largely due to superior firepower, selection forces (which take the form of killing and capturing enemy troops) are much stronger on insurgents than on regular armies. Not to put too fine a point on it, but U.S. soldiers and marines kill far more insurgents than die themselves due to insurgent attacks. But this strengthens insurgencies in the long run. On average, the weakest and least adept fighters will be killed off or captured. The tactics that didn't work will disappear. The hiding places that were easy to uncover won't be used anymore. What the survivors then do sustains the evolutionary cycle for insurgencies—they replicate their successful ideas and tactics by recruiting and training new insurgents. The net result is that the insurgency as an organization (albeit a loosely controlled one) has learned better ways to fight a regular army. What this looks like quantitatively, and has been demonstrated in Iraq, is that the ratio of insurgents killed per U.S. soldier killed is virtually unchanging, despite a huge escalation of resources and troop deployments by the United States.

This particular kind of selective learning would seem to be a good example of learning from failure, and not seemingly one we would want to replicate in our own endeavors. Learning from failure in nature usually involves death. While the unfortunate individual with the weak mutation doesn't learn, the species as a whole experiences a kind of learning by reducing the likelihood of poorly performing variations in the population. No one would advocate that we should tolerate higher mortality rates among troops because it would increase our opportunities to learn. But we do need to be aware that this type of learning can be an unintended and almost inevitable consequence of apparent short-term progress.

More helpfully, we must also recognize that learning to survive in nature is a process of learning from *both* success and failure. Learning from success reinforces mutations that benefit survival. Successes are the creative outputs that provide new working models for survival. Nature learns much more from success than from failure. A sea otter pup inherits the same particular dietary preferences as its mother because it watches her successful foraging dives and learns how to avoid sea urchin spines or crab claws. Amphibians emerged on land because using the resources abundantly available there was a much more successful strategy than fighting for survival in an increasingly crowded ocean. In fact, over the history of life on Earth, despite several mass extinctions, the number of species and the number of unique niches that they occupy have more or less continually increased.[17] The millions of extant species today, and the countless individuals of all those species, are each success stories in nature.

Societal organizations, by contrast, tend to do most of their learning from failures. After 9/11, Hurricane Katrina, and many other security disasters, long lists were made of how and why we failed to maintain security. At first blush, this would appear to be a good thing. But here, an important distinction arises between human society and the rest of the natural world. Ethically, we shouldn't plan to learn from mistakes in the way that nature does, which typically involves a lot of death. When we do learn from mistakes, it means that we are too late to prevent the suffering and loss of life that occurred in the first place.

A former student of mine, who is an active-duty lieutenant in the U.S. Coast Guard, noticed the tendency to learn from failure and ignore the lessons from success through her years of responding to oil spills and other hazards. When the Coast Guard experienced one of its biggest failures in recent years—the botched response to the relatively small 40,000 gallon M/V *Cosco Busan* oil spill on November 7, 2007, in San Francisco—it immediately set to work studying and accounting for what went wrong. The commandant of the Coast Guard ordered an "Incident Specific Preparedness Review" to be carried out by a large group of local, federal, and international agencies with expertise in oil spills and disaster response. The result, several months later, was a massive document with over 190 recommendations, a number of which were implemented in future Coast Guard practices.

Yet when the Coast Guard deftly held back and cleaned up over 9 *million* gallons of oil spilled after Hurricane Katrina and Hurricane

Rita in one of the most challenging cleanup environments possible, nary a word was spoken. In fact, the massive Townsend "After Action" report on Katrina identified 17 "Critical Challenges," 125 recommendations, and 243 action items, covering everything from search-and-rescue to transportation infrastructure to human services, but none of them addressed oil spill cleanup, the one unqualified success after Katrina.[18] The oil spilled by Katrina was one of the largest oil spills on record, approximately two-thirds the size of the 1989 *Exxon Valdez* spill. Yet so forgotten were the oil spills caused by Katrina that by the time of the 2008 presidential campaign, Republican candidate Mike Huckabee was able to argue publicly that "not one drop of oil was spilled" due to Katrina.[19]

Why would it be important to learn from the modest success of the Katrina oil cleanup when so many other aspects of the Katrina response were unmitigated disasters? The answer would come just a few years later, on April 20, 2010, when the *Deepwater Horizon* deep sea oil rig contracted by British Petroleum exploded, killing eleven people, and began hemorrhaging oil for months into the same Gulf of Mexico that was wreaked by Hurricane Katrina. Virtually none of the lessons learned from Katrina's failures would be helpful in responding to the *Deepwater Horizon* spill—the spill didn't create a refugee crises or cripple land transportation routes—but as 2.5 million gallons of oil per day poured into the gulf, it seems reasonable that previous lessons from successfully containing and cleaning large amounts of oil under extremely difficult circumstances would be valuable.

Why do we fall back on learning from mistakes rather than successes? In part, it's because we often take a business or engineering approach to problem solving. Engineers seem to take a perverse pleasure in highlighting the importance of learning from failure. There are many books on famous engineering failures, and the striking 16-millimeter footage of the Tacoma Narrows suspension bridge oscillating like a sea snake before disintegrating plays extremely well on disaster documentaries.[20] Engineers rightfully point out that these disasters help to make future bridge, building, and oil rig designs safer, and this works well in the intelligently designed world that engineers live within. But in the dynamic world of security, good designs are not nearly enough.

As with the engineers, a mantra of the business learning literature is that organizations need to learn from their failures, and a view of "failure as the ultimate teacher" prevails. A 1993 paper, in fact, praised British Petroleum as an exemplar of learning from failure,[21] noting that BP capitalizes on "constructive failure," which is defined as a failure that provides "insight, understanding, and thus an addition to the commonly held wisdom of the organization."[22] Undoubtedly, the *Deepwater Horizon* disaster provided all the components of constructive failure to BP, but it killed eleven people, saddled the company with well over $1 billion of damages,[23] was catastrophic to BP's share price, slammed the door on the permissive regulatory romper room that allowed BP and other oil companies to operate relatively cheaply in deep water, and, of course, resulted in one of the most extreme man-made environmental disasters in history. Looking to "failure as the ultimate teacher" isn't too valuable if the whole school burns down.

When we take a biological perspective on learning, we realize that we are biased toward learning from failure because of the selective forces at work. In nature, the selective agent acting on learning processes is anything that identifies one variant over another and helps it reproduce or kills it off—a violent storm that rips the weaker kelps off the rocks, a clever predator that lures deep-sea fish directly into its jaws with a glowing lantern, a picky mate that passes up the advances of any male companion whose claws or antlers or tail feathers are just a little too small.

When it comes to how we respond to big events in society, it is often news media that play the selective agent. After the *Cosco Busan* spill, images of hundreds of frustrated San Francisco volunteers waiting to clean up oiled birds, but held back by government bureaucrats, were disseminated by national media. Those kind of images result in calls to Congress and demands for investigations. By contrast, the Coast Guard's valiant attempts to clean up oil spills following Hurricane Katrina hardly made newsworthy footage relative to images of people stranded on the roofs of their flooded houses and Americans begging for deliverance from the overwhelmed refugee camp in the Superdome.

This force of selection isn't likely to change soon. We cannot (and would not want to) order media outlets to only report good news. Moreover, there are some security concerns, especially ongoing or recurring events, for which learning from the last failure is helpful. A minor computer virus embedded in an attachment that temporarily disables your computer is generally all it takes for you to scan all future attachments for viruses. But the most dramatic, most costly, and most deadly failures are often idiosyncratic confluences of events that then cause a "paradigm shift" in how we view security. As Dominic Johnson and ecologist Elizabeth Madin have pointed out, we didn't learn from earlier warning signs of these catastrophes because of a range of individual and group barriers to learning, including human psychological biases that cause us to underestimate risk or underappreciate risks that we don't sense directly, as well as institutional inertia and political disposition toward maintaining the status quo.[24] In these cases—Pearl Harbor, the 9/11 attacks, and Hurricane Katrina are all good examples—the catastrophe itself was the failure we finally learned from.

So, how can we learn adaptively when the things we are most concerned about tend to be low-probability one-time events? A now classic paper from the organizational learning literature, titled "Learning from Samples of One or Fewer," tried to address this problem.[25] The authors argue that organizations can learn from minimal experience or from less-than-catastrophic events in three ways—by deepening their experience of the event, by spreading out the experience through eliciting a wide range of people to analyze the event, or by creating hypothetical events to mimic and learn from an experience. This last approach is particularly important when considering events that could be catastrophic in reality.

Who learns like this, and does it actually work? As it turns out, human babies do, and they have an excellent track record of developing into high-functioning, highly adaptable adult humans. And babies, in turn, are not that much different than security planners or security organizations. That is, both babies and security planners typically have very little experience with the things they need to learn about. There are countless objects, situations, behaviors, and natural phenomena that are perfectly commonplace to adults that babies have never seen. Likewise, there are countless potential terrorist or cybercrime attacks, ways that fragile food distribution systems could collapse, or pathways for natural disasters to strike from.

Babies and truly adaptable organizations use the same basic methods to deal with this information deficit and make sense of an ever-changing, ever-surprising world. This is also essentially how good natural historians like Darwin make sense out of an incredibly complex and variable biological world. Darwin, for his part, didn't just dream up his theory of natural selection, nor did he have a revelation while staring at finches on the Galapagos. Rather, he spent painstaking years during and after his long voyage on the *Beagle* piecing together the variation of life. If you doubt his dedication, keep in mind just two of his lesser known works, the extensive and definitive guides to the world's barnacles: *Living*

Cirripedia, a Monograph on the Sub-class Cirripedia, with Figures of All the Species: The Lepadidæ; or, Pedunculated Cirripedes, Volume 1; and *Living Cirripedia, the Balanidæ (or Sessile Cirripedes); the Verrucidæ*, Volume 2. Not exactly world-changing books like *On the Origin of Species* or even delightful travelogues like *The Voyage of the Beagle.* But this painstaking observational study of barnacles was part of Darwin's quest to figure out how each living thing played into the larger picture of the highly variable and ever-changing world he observed during his wandering years. As he reflected in his autobiography, "My mind seems to have become a kind of machine for grinding general laws out of large collections of facts."[26]

Or, as the authors of the "Learning from Samples of One or Fewer" paper wrote, "Great organizational histories, like great novels, are written, not by first constructing interpretations of events and then filling in the details, but by first identifying the details and allowing the interpretations to emerge from them. As a result, openness to a variety of (possibly irrelevant) dimensions of experience and preference is often more valuable than a clear prior model and unambiguous objectives."[27]

The commonality of these different perspectives on an inductive approach is that learning to see the world with an adaptable mind, as babies, naturalists, and adaptable organizations are able to do, requires both a dedication to observation and a rich imagination. True, some early naturalists in the Middle Ages let their imagination get away from them at times—mixing incredibly accurate portrayals of real species with descriptions of unicorns, mermaids, and dragons.[28] Likewise, toddlers routinely talk to imaginary friends, weave elaborate tales that mix imaginary and real entities, and derive fanciful explanations for what they observe.[29]

But so do we.

The most memorable conclusion of the 9/11 Commission Report was that the security failures leading up to 9/11 represented a "failure of imagination,"[30] and this was certainly an apt description of the security situation up until 9/11. Our problem now at times seems like the opposite. We imagine too much. We see monsters under every bed, and it makes us do things, sometimes at great expense, that are fairly ridiculous. We now have no problem imagining almost anything to be a security threat, from any package over 16 ounces (which can no longer be put in a mailbox) to my daughters' yogurts, cruelly confiscated by TSA agents when we board a plane. For good reason, soldiers in Iraq and Afghanistan imagine almost any piece of roadside debris or unusual disturbance to be a threat, but many stricken with varying degrees of post-traumatic stress disorder continue to imagine these threats when they return.

How do we reconcile the active imagination needed to envision multiple states of an unknown future world with the need to spend resources on things that will truly make us safer? Babies do it by growing up. That is, babies' imaginations are better, more finely tuned, and more adaptive than the "anything goes" imagination that most adults practiced after 9/11. Their imagination is contextualized and scientific. Like good naturalists, babies mix observation and imagination to create hypotheses—educated guesses—about how the world works and how it could work. They then test multiple alternative hypotheses: they see how a toy car moves when pushed and how it breaks when it falls down the steps, how other babies react when they're touched gently or screamed at, and how adults react when they attempt to imitate different sounds—and they reject the hypotheses that don't work for them and further test those that do.

This is exactly what I do when I begin to study a new tide pool or revisit a natural location that was studied long ago by one of my predecessors. I observe with all my senses, and I compare those observations to what I think I know from reading the past scientist's notes or what I've experienced with similar species in another location. I use my imagination to tell some hypothetical stories about how this little corner of the world has changed. For example, I might find that compared to earlier studies, very few large snails still live in this tide pool; and I begin to compile some stories—a new predator came to town and ate all the large snails, warming ocean temperatures were disadvantageous to the large animals, humans began using the snails for bait, a mermaid loved the snails and brought them all to her magical undersea garden. Then I make more observations to figure out which of those hypothetical stories could be true, and which are most likely to be true. Like a baby, I am never completely correct. But I can't let that stop me. I will never have enough data or observations to be 100 percent sure that I am telling the right story, but, like the baby, I need to find the plausible story so I can move on and grow. My needs to distill this imagination and observation into a plausible story are prosaic—I might need to get the work published so I can keep my university job—or potentially a little more important—I need to identify the most likely cause of change to help a community group that wants to enact a conservation plan. But for the baby and the security organization, this ability is a matter of life and death.

How can our security organizations and strategies grow up beyond the purely imaginative stage while maintaining an ability to understand and respond to unconventional threats? Hypothetical thinking is only one of three ways that "Learning from Samples of One or Fewer" suggests that organizations can cope with uncertainty. It also suggests that making multiple experiences out of a single experience by getting many people to look at it and deconstruct it is a way to deepen that experience and learn from it. This isn't the typical way organizations work. Usually a single team is ordered from central management to analyze problems and develop solutions. This kind of centrally controlled organization was discovered to be at the root of most of the cases that Dominic Johnson and Elizabeth Madin examined in which organizations failed to learn until catastrophe struck.

By contrast, loosely related groups of individual humans, when sharing learning through networks, can adapt almost instantaneously. The best example of this is that until 9/11, the normal response to a plane hijacking was to put up no resistance: hijackers made demands that were eventually negotiable, and lethal threats were unlikely to be carried out. But within minutes of the mutation that saw terrorists starting to use passenger planes as weapons of mass destruction, humans used networked technology to share information about the change in hijackers' tactics, and passengers on one hijacked plane immediately adapted a more active defense, risking their own security to protect a larger (and largely unrelated) group of humans. Subsequent airborne attack attempts by Richard Reid and Umar Abdulmutallab were similarly stopped by passengers.

To be adaptable, organizations need to learn like babies and naturalists and networked groups of well-informed adults, not like businesses and engineers, but they're generally not set up to do so. Instead, we need to alter the genetic architecture of a security organization to be rearranged into a kind of chimera—a beast as adaptable and flexible as natural learning systems, but not so remotely different from our existing organizations that we have to destroy everything and rebuild it all. Fortunately, nature offers abundant clues on what this organization should look like. In fact, we can't look at any part of nature without seeing templates for truly adaptable organizations. The next chapter shows how decentralized organizations are extremely common in nature—in fact, it is from this kind of organization that almost all of the remarkable capacity to adapt found throughout the natural world emerges.

chapter four

ORGANIZED TO CHANGE

UNDERSTANDING HOW NATURAL security systems work requires keen observation of nature. But observing nature isn't just about peering for hours into a forest grove to catch a glimpse of a rare bird to add to your life list, or collecting butterflies to pin into a shadow box. Useful nature observation looks at how organisms interact with one another and with the world they live in. It looks at how these interactions change through time—over a matter of minutes, through the seasons, and across eons of geologic time. And good observation engages all of the senses.

One of the best observers of nature I have ever met is paleontologist Geerat Vermeij. He can identify nearly any fossilized or living marine organism, and by carefully observing the shell or carapaces of any individual, he can tell you how it lived, how it died, and how it changed relative to its ancestors. Even the most seasoned naturalists are amazed by Geerat's skills in the field. His abilities as a naturalist are even more impressive when you take into account that he has been blind since the age of three.

But just as his lack of eyesight hasn't hampered his ability to observe the natural world, it hasn't restricted his vision in connecting the pieces into a larger whole. In this he follows the tradition of Darwin, Ricketts, and Rachel Carson, who built grand holistic syntheses based on meticulous observation of the variation in nature. Vermeij uses his incredible tactile sense to build, in his mind, an immense catalog of biological variations from which emerges a holistic picture of patterns and trends in the evolution of life. Twenty years ago, Vermeij published one of the first treatments of arms races in the natural world in his book *Evolution and Escalation,* in which he analyzed millions of years of battles between snails episodically growing in bursts of stronger and more armored shells as their various crab predators developed ever more powerful claws.[1]

As Vermeij continued to observe the diversity of life recorded in fossils that span hundreds of millions of years, he gradually came to understand a prominent pattern of nature, one that has critical importance to the study of natural security systems. He found that for the most part, centrally controlled organizations do not thrive in nature. Rather, the job of sensing the environment is farmed out to multiple agents that have a great deal of power to respond on behalf of the larger organism. As with all of the natural security systems I discuss in this book, this decentralized pattern of organization occurs at every level of biological structure. Cells, individuals, and even ecosystems survive and adapt by letting localized agents detect environmental change and respond to it with little central control.

Starting small, the vertebrate immune system is a wonderful example of an adaptable organization. Although it serves a centralized purpose (to keep the body free of malicious invaders) it works by sending out multiple independent cells that identify invading organisms. These cells don't have a predetermined "watch list" of intercellular terrorists handed down from the brain for them to check against. Rather, they map out the characters of anything entering the body and then call in helper agents to bind to the invaders and neutralizing agents to destroy them, if necessary. All of this happens continuously and without a lot of preplanning. Of course, past experience with certain invaders improves performance of this system, which is why vaccination using small doses of pathogens is so effective.

Like the immune system, our brains have to deal with a huge array of uncertain information. In the previous chapter I mentioned that one way the brain anticipates what will come of all this uncertainty is by organizing all past information into relational chunks. The brain does this with a highly decentralized storage system that places different aspects of memories (the smells, the colors, the emotional feeling associated with it) in different places. The ensuing network of related neurons is robust—the memory can't be removed just by an injury to a particular part of the brain—but also flexible and adaptable.[2] It's been argued that at a higher level, human intelligence generally mirrors the intelligence system of a single brain—that is, our collective intelligence as a species is decentralized, but still a robust collection of knowledge, experience, and learning linked by social networks.[3]

Back at the level of whole individual organisms, one of the great advancement of birds over their dinosaur ancestors was decoupling bones connecting the legs and tail, which were essentially fused as a single locomotive unit in dinosaurs. The specialization allowed independent movement of the tail, which provided key stabilization to wings for a new life of living in the skies.[4] Other organisms develop adaptive organizations by specializing individual clones. That is, species like corals, tunicates, and bryozoans start as genetically identical clones, dividing asexually from a single parent cell. As the colony smears out through division across a hard surface such as a submerged boulder or the skeletons of long dead corals, each unit can take on an important function—some becoming specialized for feeding, some for reproduction, and others along the edge of the colony armed with stinging cells to repel other colonies and gain precious space on the hard surface—and all contributing to the overall survival of the colony.

Decentralized organization doesn't just benefit individuals. Populations of individuals also utilize this system of organization. For example, schools of fish maintain safety for the individual fish within them by independently responding to changing ocean conditions. No supreme leader fish tells all the others what to do—rather, each fish doing its own part to sense change and change itself leads to the appearance of a much larger whole organism.

Simon Levin of Princeton, who has built a career brilliantly uncovering linkages between the properties of mathematical functions, economic game theory, and ecosystem and societal organization, finds that this same decentralized organization is what gives ecosystems resiliency in the face of environmental change.[5] Levin isn't saying that ecosystems actually evolve as a unit, but rather that the individual organisms within the ecosystem—each doing their own thing to survive—impart survivability on the ecosystem as a whole. This turns our whole notion of a hierarchy on its head. We tend to think of hierarchies as controlled from the top, exerting power down through the system. Ed Ricketts saw this reversed hierarchy as a basic property of biological organization: "Each higher order, instead of ruling the ranks of individuals below, is actually ruled by them. Each rank is completely at the mercy of its subjects, dependent on their abundance or accessibility."[6]

Levin sees the same patterns in social and political patterns that emerge when people in adjacent areas tend to have similar belief systems and vote in similar patterns. Levin calls the systems that emerge—whether ecosystems or voting blocs—"complex adaptive systems," and their most remarkable characteristic is that they emerge from very simple and largely independent actions of individuals within the system.

Decentralized and distributed organizational systems are adaptable for three main reasons. First, multiple sensors all looking or experiencing the environment from their own perspective provide more opportunities to identify unusual changes and unexploited opportunities. When we let a single entity (say, TSA at airports) take complete charge of security, the number of observers goes down, along with the probability of identifying a threat to security. Second, multiple agents committed to the security mission in their own local area create opportunities to specialize tasks, so that energy isn't wasted in having every part of the organism doing the same things; rather, those doing the most important things (e.g., providing defense when hostile enemies are around or reproducing when populations are low) get the resources to replicate their activities. We have often ignored this lesson in distributing resources for homeland security. Recently, governors throughout the United States were appalled to find that in order to receive federal funding from the Department of Homeland Security, they had to commit 25 percent of their budgets to defense against improvised explosive devices[7]—a huge threat in foreign conflicts but extremely low in importance relative to other threats facing the states. Third, distributed sensors respond to the most immediate environmental conditions in time and space—they see the environment for what it "is" rather than what it "should" be according to some preconceived notion. This way, the octopus that gets transferred to a lab tank isn't paralyzed by the new environment. It simply uses its eight tentacles and thousands of suckers (which can smell, by the way) to feel out its new surroundings, search for food, and find an escape route.

To appreciate how vital these organizations are to adaptable security, consider what happens when we fail to use them.

THE FIRST POST-9/11 TEST OF HOMELAND SECURITY

August 29, 2005, started like any other day in post-9/11 America. On that Monday, roughly 1.8 million airline passengers were busy taking off 3.6 million shoes and putting them through X-ray machines, going through the security ritual necessary before boarding their flights. Nearly four years after 9/11 and then Richard Reid's attempt to blow up an airplane with a shoe bomb, passengers were getting used to long lines and invasive searches at airports. The multi-billion-dollar Transportation Security Administration (TSA), part of our newly created Department of Homeland Security (DHS), could proudly say that there was virtually no danger from a shoe bomber on August 29, 2005. But by the end of the day, something else had happened: an entire U.S. city was underwater. Families were separated, and senior citizens were drowning as floodwaters rose and no one was there to save them. Chaos and disease were festering in the shattered shell of the New Orleans Superdome, once considered to be an engineering wonder of the modern world, now a makeshift shelter. From an ecosystem perspective, the destruction caused by Hurricane Katrina was inevitable. Just as the hardest hit areas of the Asian tsunami were those where protective mangrove forests had been removed for coastal development and aquaculture ponds, New Orleans, a city already built below sea level behind levee walls, had stripped most of its protective wetlands for its own economic development. Like ignoring animals that are running uphill before a tidal wave, actively destroying natural protective systems is an obvious failure of our security responses to respect nature's experience.

But Hurricane Katrina also stripped bare the more subtle and less appreciated ways in which we fail to capitalize on free security strategies from nature. Much of the trauma to New Orleans residents occurred after the storm passed, when it became clear that the emergency response systems we had been obsessively developing and reorganizing after 9/11 were not working at all. One of the most repeated questions in the aftermath of Katrina was "Where was FEMA?" referring to the Federal Emergency Management Agency, which was ostensibly in charge of disaster relief efforts. Because it is a bureaucracy, the best way to find FEMA is by looking at the "org" chart of its parent organization, the Department of Homeland Security (see Figure 2, page 70).

When you finally find FEMA (Hint: it is the box with the dashed outline) you can see that it is literally buried in a huge stack of blocks, all representing their own enormous bureaucracies—such as the Coast Guard and TSA—all required to run decisions up the chain of command to the Secretary of Homeland Security, and, consequently, all vying for the secretary's attention.

An organization like this might work fine in carrying out a planned set of tasks that continue routinely day after day. It's like an early circuit board with a finite number of pathways through which the energy of decision-making can pass. But as I've argued, security problems are security problems precisely because they are not routine—they are highly variable and unpredictable. If one of the organizations inside one of those boxes needs to do something completely different than normal—as FEMA needed to after Katrina—it has little recourse to do so.

That's not to say that some organizations didn't demonstrate some amazing responses to Katrina. The U.S. Maritime Administration, a branch of the Department of Transportation that maintains a fleet of ships as well as contracts with multiple vessel owners to make vessels available during wars and national emergencies, quickly set up shipboard spaces that the various security agencies used as command centers.[8] And, of course, many individuals within all of the agencies, as well as individual citizens, improvised all sorts of effective responses to the hurricane. These were exceptional cases, and like the adaptable soldiers in Iraq and

Figure 2: Organizational chart of the Department of Homeland Security

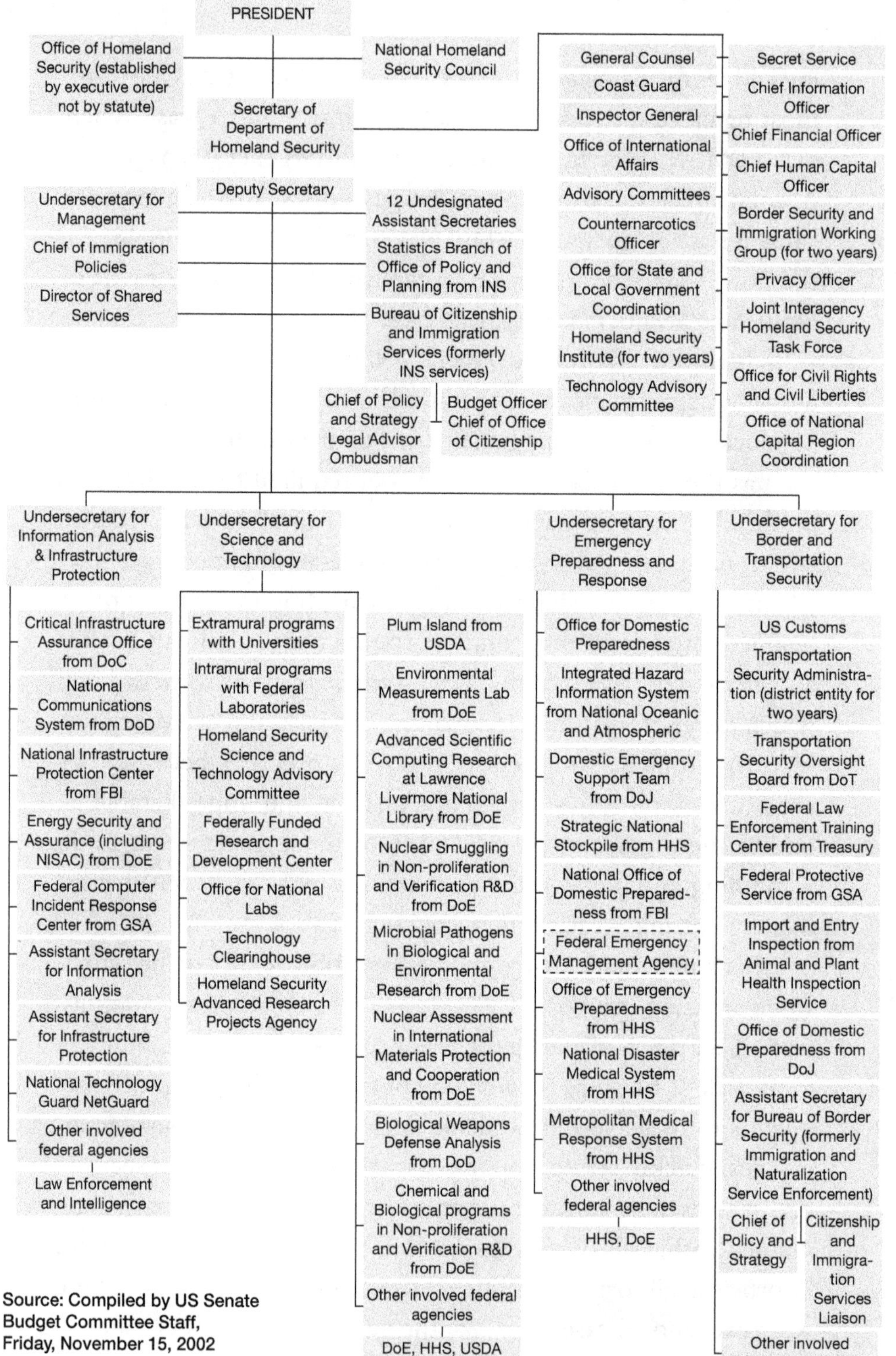

the passengers on United Flight 93, they learned quickly to do the best they could with the resources they had. What we would like to know is, how were they able to adapt so quickly, and is their adaptability something we can replicate without having to wait for the next disaster to bring it out?

PIONEERS OF DECENTRALIZED ORGANIZATION

It is often assumed that the stack of boxes leading to one central controller is the natural and inevitable way an organization develops. And people working within such an organization often assume that there is no way to change that system of organization without destroying the entire organization itself. The first assumption is, in fact, completely false, as proven by most successful biological organizations on Earth. And challenging the second assumption, which is beginning to happen in societal organizations throughout the world, is the key to turning nonadaptable *organizations* like the DHS, as well as the Department of Defense, discussed in the prologue, into adaptive *organisms* that learn from changes in the environment, react quickly to them, and ultimately keep us safer. Fortunately, pioneering individuals and organizations around the world are providing us living proof of how adaptable we can be, even in a world of stacked boxes and organizational charts.

Captain Douglas Cullins, a straight-talking U.S. Marine who from an outside perspective would appear to be some kind of new hybrid fighting organism—combining the stock-straight posture, reverence for tradition, and win-at-all-costs attitude of a good marine with an obsessive desire to continually adapt to whatever is thrown his way on the battlefield—recently briefed a group of academic scientists and military analysts on what he and his marines did to adapt to a combat environment in Ramadi, Iraq, that was far different than what any marine had seen before.

Cullins and his marines didn't adapt immediately, but over the course of several tours that saw every kind of fighting and vacillated from infuriatingly quiet to shockingly violent, they realized that they were fighting a war they hadn't been perfectly trained for. While they knew that their "kinetic" (now termed *lethal*) fighting ability—the firepower and combat training they all possessed—was their best asset, it became clear that Iraq called for something completely different: electricity, water infrastructure, schools, road projects, and safe places for government meetings. Becoming a marine company that provided these things was especially hard for eighteen- and nineteen-year-old warfighters who may have just seen their best buddy killed by a roadside bomb to accept, so Captain Cullins laid out their options in terms they could clearly understand. "You've got to understand that marines, more than anything, want to win," Cullins says, "so I let my marines know that if we were going to win this kind of battle, we had to adapt."[9] Part of this adaptation was simple posture—walking through town in a relaxed way, with guns shouldered and impact glasses down so people could see the eyes of the Americans. Most adaptations undertaken by Cullins's company had little to do with fighting and everything to do with learning about and understanding a different culture's modes of communication and specific needs. Cullins completely reorganized his company's structure and modified its mission to suit the environment. His heavy weapons platoons and sniper teams were given tasks that, in addition to providing security and hunting down insurgents, included a deliberate mapping of Ramadi's essential services infrastructure. Blown power transformers, clogged sewers, and empty fuel storage sites quickly became critical information. His lieutenants and senior enlisted were given great latitude in identifying problems and implementing solutions. Beg, borrow, "acquire," or leverage, his marines were given the mission to get the city up and running.

The results came quickly once the marines made the shift in mindset. They quickly "leveraged" the local populace's expertise and spent countless hours repairing the damage caused by years of brutal fighting. Most civilians did so willingly; the few holdouts were subjected to a low-level "shame and humiliation campaign" designed to let the local populace know which of their leaders were refusing to fix the problems, with a picture and home address of the holdout posted on street corners. Cullins quickly learned that rival or peer pressure was often the most effective approach when trying to make a point with this population. I asked Captain Cullins if he inquired above his rank for permission to breach military protocol in this way. He chuckled slightly and conspiratorially admitted, "No, we kind of learned not to ask for permission until after a good idea worked out." He continued, "The trick was modifying the marines' mindset while retaining a simple rule that they could appreciate: be polite, be professional, but have a plan to kill everyone you meet."

Before Hollywood gets any ideas, I should note that Cullins wasn't some renegade commander, brashly defying orders to triumph in his small corner of the war. U.S. Marines are trained to operate with mission-type orders and given an end-state, termed the Commander's Intent; how subordinates accomplish the mission is up to them. Indeed, most marine companies in Iraq were adapting in similar ways—using the observational skills of their marines to determine that the battlefield was much different than the one they were told would exist, trying out a wide variety of solutions, and sharing their stories of success and failure with other companies via satellite calls and e-mail. They didn't wait for the next edition of the Department of Defense counter-insurgency manual to be revised, printed, and handed out; they just adapted. Captain Cullins, who was promoted to the rank of major soon after the conference, is currently serving as a faculty advisor at the Marine Corps Expeditionary Warfare School. There he has the opportunity to teach future company commanders his particular form of operational adaptation.

But can the actions of a few clever individuals scale up to the level of the organization itself? As a biologist who has witnessed the nested quality of adaptable biological systems, I am convinced that such scaling up is the natural inclination of organizations. Practically every living organism is a scaled-up version of an early working prototype molecule that was driven to reproduce itself. We ourselves are also scaled-up success stories, whose every cell contains a copy of that early four-letter narrative form known as DNA.

Although I've highlighted failures to scale-up adaptability in our institutions, there are in fact many emerging examples of successful, decentralized models of organization. Small bottom-up organizations around the world are rapidly becoming far more effective at promoting environment protection and social justice than the huge centralized and much better funded nongovernmental organizations (NGOs) and governments.[10] Businessman and social activist Paul Hawken likens this growing network of local organizations to an immune system, in that it is widely distributed yet connected, and grows larger not for its own sake but through the process of local populations seeing additional needs and replicating their successful efforts.[11]

Decentralized organizations reveal how truly adaptable they are when pitted against an organization still tenaciously holding on to the old model of central control. In their business book *The Starfish and the Spider,* Ori Brafman and Rod Beckstrom chronicle the struggles of traditional centralized business models in competing with new distributed networks of competitors. For example, highly decentralized music file sharing networks—typically run by college students on hundreds or thousands of independent machines that are constantly changing—have been shockingly successful at eluding the copyright protection efforts of the much better funded and highly centralized Recording Industry Association of America.[12]

Infectious disease is a security problem ripe for decentralized solutions. Emerging diseases often strike first far from medical facilities and far from large centralized aid organizations. They tend to hit poor communities with few monetary or medical resources. They also tend to mutate rapidly, meaning that solutions need to be custom tailored to the disease and the environmental conditions in which it arises. Early detection of emerging infectious diseases is essential. But it can't be done from the Centers for Disease Control headquarters in Atlanta or the UN headquarters in New York. These diseases turn deadly to humans when a virus from a not distantly related mammal mutates so that it can infect people. The first point of contact between the newly mutated disease and humans is often in places where people still hunt daily for a living. The close contact with the freshly killed animal provides ample opportunity for a still-living virus to leap into its newly available human host. This is why Nathan Wolfe's plan for stopping the spread of new infectious diseases doesn't involve a team of doctors mapping outbreaks from afar. By the time new diseases are visible to most health programs, and they marshal the resources to respond, it's too late. Rather, through Wolfe's Global Viral Forecasting Initiative he turns the hunters into their own disease monitors, having them collect blood samples in the field and report outbreaks to the forecasting network.[13] Other networks of disease spotters are emerging using online wiki tools that allow users to add sites of disease outbreaks to online maps. Even those without an Internet connection can report to disease networks with cell phones, which are increasingly common even in poor and remote locations.[14]

These strategies don't just apply to raggedy bands of activists or altruistic health practitioners. Internet giant Google has thrived on a decentralized model in which even low-level managers are given a high degree of autonomy to develop new ideas[15] and billions of users on the Internet are incorporated as selective agents to test the ideas. Google's products themselves reflect the value of independent sensors. For years now, Google has been quietly testing Google Flu Trends, an application that tracks users' "Googling" words associated with the flu, such as "flu symptoms," "flu remedies," "flu vaccines," and so forth. These billions of users show clear patterns of search that are well correlated with actual flu outbreaks. By looking retrospectively over the past four years, Google analysts found that Google Flu perfectly matched the Centers for Disease Control and Prevention's (CDC) official flu tracking, which is based on doctor and hospital survey data returned to the CDC, with one major difference. The decentralized Google Flu provided data on outbreaks two weeks faster than the centralized CDC data.[16] When it comes to the rapidly mutating flu virus, two weeks' advance notice could easily be the difference between a mild nuisance and a global pandemic. And should a pandemic occur, it's worth noting that most of the 40–100 million deaths from the 1918 Spanish Flu pandemic occurred in a sixteen-week period, with up to a million deaths per week worldwide and 10,000 deaths per week within individual U.S. cities[17]—every week counts.

FEAR OF A DECENTRALIZED PLANET

The successes of octopuses and coral reefs, law-breaking music file sharers, and business giants like Google can all be replicated at every level of security analysis, planning, and practice in society. When discussing this radical idea of decentralized organization with people ensconced in highly centralized organizations, I try to reassure them with the dictum "Have an open mind, but not so open your brain falls out."[18] That is to say, I'm not advocating a completely independent set of decision makers randomly going around doing different things. For all its decentralized camouflage and shape shifting, an octopus is still an octopus. An immune system still serves a body that is identifiable as a given species. And Google is still a company with a corporate image and a goal to make profits.

To be more blunt, before we rip up our org charts and depose all our leaders, all those highly distributed sensors and rapid responders do pretty poorly when left completely to their own devices. Decentralized agents will provide a lot of what is needed, but three critical ingredients must be added to a completely independent network to make it an adaptable organization. First, it must be built around a challenge. In natural systems, that challenge is simply to survive and reproduce. There are likely to be sub-challenges beneath that, such as to live through a long cold winter, or to find water during a drought. The challenge takes all the varied energy of the independent agents and provides direction. Imagine a centipede where each individual leg just went wherever it wanted, whenever it wanted—it would be pretty comical, but not very well adapted. Second, it needs resources that are made available to the decentralized units to solve the problem. In animals, this could be a body to safely house all those nerve cells, or a mouth to feed them, or an alpha pack leader among social animals that organizes their hunting. For human groups, money is always nice, but sometimes the necessary resource is recognition, publicity, prestige, or career advancement. Sometimes it is a trusted place, like a well-respected foundation or well-known organization, that will attract and store resources that can be then distributed to the most adaptable and successful agents. Third, there needs to be some filter on all the information that comes back from many independent sensors. Even a fabulous array of astute sensors may bring back a lot of information that isn't very relevant to our needs and end up wasting just as many resources as a stodgy, completely centrally controlled bureaucracy. Scientists have recently discovered immune system molecules, unexpectedly, in the delicate neural architecture of mammalian brains. Elsewhere in the body, these "search and destroy" molecules run around freely to identify and neutralize invaders. In the brain, however, these molecules apparently serve as regulators to prevent too many neural connections from being made, which would overwhelm our sensory systems.[19] Lack of these regulators may explain the sensory overload experienced by those with schizophrenia and autism. So, in all systems some selective mechanism is needed—a way to rate and prioritize information coming in; this is best done when it relates directly to the central challenge.

After 9/11, al-Qaeda, though widely the object of hatred, was also respected for its organizational structure. Security analysts noted how it functioned as a semi-autonomous network of cells that could quickly shed units without compromising the whole organization. All of the 9/11 hijackers, for example, had very little contact with one another, although they shared a common mission and received resources from the central organization.[20]

More recently, however, al-Qaeda has lost its effectiveness. Marc Sageman, who has studied terrorist networks from the ground up, carefully mapping their components, relationships, and changes, has shown that swinging or being pushed too far toward a decentralized system can greatly reduce the influence of an organization. He has found that within the past few years, with key al-Qaeda leaders killed, captured, or pursued into hiding, the movement has evolved into what he calls a "leaderless jihad," still wildly popular with disaffected Islamic youth looking to fight a heroic battle with the West, but far less capable of carrying out terrorist actions. As Sageman notes:

> Without the possibility of a physical presence or ability to negotiate with its enemies, the social movement is condemned to stay a leaderless jihad, an aspiration, but not a physical reality. Finally, al Qaeda has not been able to forge any state allies that might protect it against Western aggression. Without a viable and effective sanctuary, it cannot fully regroup and consolidate into a physical power capable of capturing some territory in order to establish its utopia.[21]

Unlike decentralized music file-sharing networks that can actually cause damage to the large centralized recording industry (because they trade in exactly the same resource), the small cells of the leaderless jihad are not a continual threat to organized nations. But they are also like file-sharing networks in that there is a fairly low cost to enter cells of what Sageman calls "wannabe" terrorists on the Internet. There is a concern that these cells will gain prestige and converts as long as they are seen as defending Islam from an existential threat from the West. According to research by Sageman and anthropologist Scott Atran, young people attracted to the leaderless jihad are looking for opportunities to be heroes in a global cultural clash between Islam and the West. Any signs that indicate the West is on a global crusade will reinforce the sense of jihad as a heroic cause. The fear is that given enough reinforcement of their message and enough resources, this leaderless jihad could re-form into the kind of dangerous distributed network that al-Qaeda was on September 11, 2001.

But the "war on terror" environment that first splintered al-Qaeda and then, through the invasion of Iraq, helped convince young recruits to join the virtual jihad has changed rapidly and continues to do so. U.S. combat troops are pulling out of Iraq, and Muslim-on-Muslim violence has become the predominant theme there. The multiculturalism and freedom of religion offered in the United States differs markedly from the hostility perceived against Muslims in much of Europe (France's ban on Islamic veils, for example), and Muslims continue to see the United States as a land of opportunity. All of this diffuses the message that had originally united al-Qaeda as an adaptable, decentralized organization.

In other words, although the leaderless jihad has plenty of distributed agents spreading hatred on the Internet, it still lacks both the resources and the central challenge to make it a truly adaptable organization.

Adaptable organizations combine the resources, power, and unified goals of a central controller with the nimble and adaptive actions of multiple semi-independent sensors. Such systems evolved in nature over millions of years. We don't have such temporal luxury. Can such a system be created out of a completely decentralized system? Can it be created out of a completely centralized system?

STANLEY VS. THE OSPREY: A BATTLE FOR THE ADAPTABLE SOUL OF A BUREAUCRACY

Two recent weapons development programs, both developed out of the U.S. Department of Defense, will illustrate how the same organization can be woefully inadaptable—wasting time, money, and human lives with little to show for it—or spectacularly adaptable and innovative. They illustrate the contrast between centrally controlled, "intelligently designed" systems shielded from selection, and systems that emerge, as in nature, by independent problem solvers under the sort of pressure that one also sees in natural selection.

The V-22 Osprey is an odd chimera of a military vehicle, part helicopter, part airplane.[22] It was dreamed up by military planners envisioning the perfect aircraft following the failed attempt to rescue American hostages in Iran. They ordered it to have the ability to be able to swoop agilely into hot landing zones, drop off or remove troops, then convert into an airplane and fly away fast. Unlike an airplane, which is likely to crash if the engines fail, the Osprey was to have a helicopter's ability to "auto-rotate" with the rotors spinning freely, to a survivable landing in case of engine failure. Soldiers or marines would be able to exit quickly from low altitudes via static lines, or leave from a rear ramp upon landing. The Osprey was to have air-combat-worthy maneuverability and a massive forward facing canon to ward off enemy fire. It was to be protected from nuclear, chemical, and biological fallout.[23]

The program to build the Osprey began in 1981. The first Ospreys ready for combat were deployed in late 2007, twenty-six years after a contract was awarded to the only entrant in the competition to produce them. In 1986 it was estimated that 1,000 Osprey would be produced in ten years at a cost of $37.7 million each. The 1996 deadline was missed, but a year later a grand total of five Ospreys were delivered to the U.S. Marine Corps. By the end of 2009, only 181 had been procured, at a cost of $93.4 million each.

We've come to somehow accept massive cost overruns and ridiculously unmet timetables in military procurement contracts, as if it's just business as usual for the Pentagon. What's more unusual and disturbing, though, is how the adaptability of the Osprey suffered during this long development time. Unlike biological organisms, which when given adequate resources become better adapted through time (or else go extinct), the Osprey became both *less* well-adapted through the course of twenty-six years of development and *less* prone to extinction, raking in $25 billion in funding and an eager gaggle of senators and representatives ensuring its ongoing existence.

"Hot" landing zones are places where enemy fire threatens deployment and evacuation of troops, but the Osprey lost its forward-facing canon during development, leaving it without any ability to provide forward covering fire as it comes in to land. In fact, by the time of deployment, its only armament was a machine gun that could only be operated by a soldier strapped to a safety harness and firing out of the back of the aircraft with the rear cargo door wide open. The complex aerodynamics of the tilt-rotor design also stripped the Osprey of the ability for its huge rotors to "auto-rotate" in the case of engine failure, which would allow for a hard, but survivable, landing. Those same rotors do, however, create such enormous downdrafts that they melt through the decks of navy vessels at sea,[24] and in the desert landscapes where they operate now they produce dust storms that make static line drops of soldiers from the air virtually impossible. The whole design creates such instability that pilots need to drastically slow their landing speeds, making them at least as vulnerable as a normal helicopter. Given the complex aerodynamics, twenty years into the program the U.S. Navy eliminated the requirement of "air combat" maneuverability, requiring the Osprey to only have the much lower level of "defensive maneuvering" capabilities. The navy also scrapped the planned protections against nuclear, chemical, and biological fallout. Unlike a helicopter, the Osprey has virtually no visibility, what a critical Pentagon report called "situational awareness," in the troop compartment, so the independent sensors normally provided by passengers just aren't there. Crew chiefs, who on a normal helicopter are able to provide information to the pilots about the threat environment, reported that the lack of visibility from an Osprey is its biggest weakness. Finally, the Osprey performs more poorly than any other existing search-and-rescue helicopter at high altitudes, and it's actually unclear if it can perform its missions at all above 2,000 feet. Altitude wasn't a problem in Iran, nor was ice, but both are concerns in Afghanistan, and the Osprey can't handle them.

Ultimately, the combination of these little compromises led to a near complete compromise in the Osprey's originally designed mission. In Iraq, Ospreys were confined to the "low threat theater of operations"—basically they ferried troops and equipment around. Even in this relatively tame mission, the Osprey was given the handicapped goal of being "mission capable" 82–87 percent of the time. It failed. At best, in the first year and a half of operations in Iraq, the Osprey was mission capable 68 percent of the time. In part this low reliability was due to constant parts failures. Even though enough parts were sent to supply three times the number of operational Osprey, the surplus of parts was burned through quickly. Several key parts lasted only 10 percent of their expected lifetime, and others could take up to a month to replace. As with natural organisms under extreme resource stress, Osprey crews took to "cannibalizing" other Ospreys for parts.

At some point, even the Osprey's congressional shield of invincibility shattered. After hearing top military brass testify about the supposed successes of the Osprey and independent military analysts give more sober reports—including an analysis showing that an Osprey rescuing twenty-four people from an embassy could travel only 60 miles before needing refueling, compared to 400 miles for a normal helicopter—Rep. Edolphus Towns, chairman of the House Oversight and Government Reform Committee, called for the extinction of the Osprey, saying, "It's time to put Osprey out of its misery."

By contrast, another weapon borne out of the military bureaucracy, "Stanley," is a modified Volkswagen SUV filled with jury-rigged cameras, radars, motion detectors, and erector-set–like controls of its steering and gear shifting that can safely navigate through complex environments it has never seen before without any human intervention. Stanley was the best of several successful solutions for the challenge of creating a fully autonomous military vehicle. It cost only a few million dollars to produce and was completed just a few years after the initial call for designs.

It would be completely unfair to compare Stanley and the Osprey in head-to-head competition. Producing a combat-ready aircraft that can fly like both a helicopter and an airplane is a much more complex mission than making an autonomous terrestrial vehicle. But in making this comparison, I am much more concerned with the *process* of how the weapon systems were created than the end products themselves.

Unlike the Osprey, Stanley was built in response to an open public call by the Defense Advanced Research Projects Agency (DARPA) for designs to solve a simple and clearly stated "Grand Challenge." There was a modest $1 million prize offered for the wining entrant, barely enough to cover the likely costs of development, and nowhere near the billions of dollars assured to the contractor for the Osprey. Nonetheless, this call resulted in over 100 entries, mostly from university engineering departments motivated by the professional challenge and the potential prestige that the winner would accrue.

But prestige was hardly the buzzword surrounding the first entrants into the Grand Challenge. In the first year, all of the competing entrants failed their task. News reports tinged with bemused mockery outlined the various ways in which different vehicles veered off course, accelerated full force into large objects, caught on fire, or just sat motionless at the starting line, overwhelmed with possibility.[25]

Fortunately, instead of labeling the challenge a failure, DARPA simply ran the challenge again, even upping the prize to $2 million the next year. Entrants in this second Grand Challenge learned from the big mistakes and the small successes of the first generation. Repeat challengers went back to their labs to eliminate the former and replicate the latter. Teams also learned through interaction with the other teams—sharing information and surreptitiously cribbing good ideas. Whole new ideas emerged from new entrants. By the second year, these independent problem solvers produced a number of successful designs, including Stanford University's Stanley, an inappropriately blue-painted Volkswagen Touareg (a close match to Stanford's school color, cardinal red, was already taken by another team) that tore up the course faster than any other team. DARPA then kept the adaptational pathway open by offering more complex challenges.

A famous old Volkswagen ad juxtaposes Stanley's ancestor, the Beetle, with the original Apollo lunar lander under the tagline, "It's ugly but it gets you there." This slogan could apply to that ugly ocean sunfish, the *Mola mola,* and countless other species in nature. Good designs in nature aren't designed at all. Rather, they use an adaptable organizational structure and some feedback from nature to solve the problem at hand. Switching our security strategy from one that designs solutions to one that continually adapts solutions as the environment changes is a radical departure, but one that is already being adopted.

How do we do this? It is ironic that the lesson to be learned from this military operation is to move away from giving orders and toward providing challenges. Orders tend to assume there is one solution to a problem that everyone must follow to solve the problem. Challenges are posed with the assumption that there are many potential solutions to a problem and that the more people involved, the more likely we are to find a really outstanding solution.

Challenges essentially create the same three features of adaptable organizations seen in the natural systems. They bring in multiple problem solvers with diverse skills and perspectives. Because respondents to challenges are small localized groups and individuals, challenges encourage these multiple agents to work on very specialized problems and sub-problems as they arise, rather than trying to solve a much larger conglomerated security issue. And finally, because challenges exert little central control beyond the initial problem statement, preconceived notions of how things are done are essentially thrown to the wind with the challenge statement itself. Participants can use any method they think appropriate to solve the challenge.

There is no reason the DARPA Grand Challenge model can't be adopted for other societal challenges. In fact, it's already happening. As a marine ecologist, I have spent enough time with fishermen to know that they are deeply skeptical of government agencies trying to regulate their livelihoods—constantly ordering the fishermen to do different things and use different gear to reduce their environmental impact. They rightly believe that as the people who continually work on the ocean, they know considerably more about how best to fish than bureaucrats and their computer models. Nonetheless, even fishermen acknowledge that bycatch (all the species accidentally caught in fishing gear that can't be sold as part of the main target catch) is a huge problem that costs fishermen and hurts the environment. Rather than waiting for a government agency to order fishermen to change their gear or their practices to reduce bycatch, the World Wildlife Fund proposed a challenge—open to all fishermen, environmentalists, backyard inventors, really anyone—to come up with better bycatch reduction devices that could be cheaply and easily used by fishermen. With a small but significant $30,000 cash prize and a chance to take control of their fishing efforts rather than be told what to do, many fishermen (as well as backyard inventors and concerned citizens) jumped at the challenge, which has now been repeated for several years running,[26] with several successful innovations appearing each year.

Challenges have also appeared recently to solve long-standing mathematical challenges, to figure out why Toyota cars were suddenly accelerating,[27] to develop better ways to clean up oil from the *Deepwater Horizon* blowout,[28] and to reunite a found camera with its owner.[29]

I would be remiss to close this chapter without acknowledging the one unqualified success in the development of the Osprey. This was an odd political mutation of the Osprey known as the Congressional Tilt-rotor Technology Coalition, whose members cajoled and badgered their congressional colleagues ceaselessly to support ongoing funding of the Osprey program. Not surprisingly, it stems from the one truly decentralized part of the aircraft's developmental process, which was that its component parts weren't made in one huge Bell helicopter factory, but by over 2,000 subcontractors in over forty states.[30] While this didn't necessarily improve the survivability of an individual Osprey under enemy fire or harsh environmental conditions, it did ensure the program's survivability in Congress. Because so many congressional districts had jobs making parts of the Osprey, there was always a congressperson ready to defend the Osprey every time a selective budget axe appeared.

As a former congressional staffer, I don't get too judgmental about this stuff; I've long ago accepted the selective forces on congresspeople and how it drives their own adaptation—they need to bring jobs and money to their districts to appease the selective forces of voters in November. As a biologist, what I find fascinating about this aspect of the Osprey story is how doling out congressional favors aided the survival of this dodo. It did so by providing many layers of *redundancy* in the system (if one congressperson didn't stand up to defend it, there were many others at the ready, and when those others were able to form a group—as they did with the Tilt-rotor Coalition—they became both redundant and robust). Redundancy gets a bad rap in our society, but it's well embraced in nature for good reason. The next chapter reexamines redundancy—an often maligned concept—as a deeply biological capacity that builds flexibility, creativity, and ultimately, survivability.

chapter five

NECESSARY REDUNDANCY

We tend to think of redundancy as wasteful. The path of industrial progress toward ever greater "efficiency" mercilessly searches for the fastest route from raw materials to marketable product, and that route should never circle back on itself or visit the same place more than the minimum number of times necessary to create the product. Politicians who want to burnish their fiscally responsible credentials bark that they will "go up to [insert relevant capital here] and trim the fat!" And the latest corporate term for eliminating jobs, "rightsizing," tries to intimate that a company with the previous higher number of people must necessarily be the wrong size. The people who are fired from their jobs as a result of all this efficiency, fat trimming, and rightsizing are said to have been "made redundant."

If redundancy is so wasteful, why is it so common in the natural world, where organisms are constantly struggling to obtain even the minimum resources needed to survive? And conversely, if redundancy is so common in nature, which is exceedingly adaptive, does getting rid of redundancy in society make us less adaptable? The answers lie in *how* nature uses redundancy.

There is in nature lots of plain old repetitive redundancy. DNA typically contains the code for multiple copies of important genes, which protects against damage to that section of the DNA molecule and serves as a hedge against errors in transcribing the genetic code during cell division. Species like centipedes and millipedes have multiple legs all working on the same issue of locomotion. Many species produce huge numbers of offspring under the probabilistic assumption that only a few will survive to reach their reproductive years.

This basic pattern of repetition is what Geerat Vermeij calls "unimaginative redundancy."[1] There's a safety in numbers to having a lot of legs or a lot of babies, but if they are all doing the same job, or are just thrown out into the world with little chance for survival, they can't do much independent problem solving, the stuff that advances you evolutionarily. But sometimes this is enough. Certainly, just a little bit of repetitive redundancy in something like the financial system would have gone a long way after 9/11—on that day virtually all of the trading infrastructure was held in the World Trade Center towers, and financial markets were crippled for weeks after their collapse. Accordingly, one of the first adaptations of the financial sector after 9/11 was to create simple copies of their cyber infrastructure in multiple locations.

Consider the value of redundancy to an insurgent force or the army of a small country. With little financial resources, the mega weapons held by the superpowers they may be fighting, like aircraft carriers, long-range missiles, and even fighter jets and attack helicopters are well out of range of their financial capabilities. Yet insurgencies have been remarkably successful in fighting much stronger forces, and their relative success has been improving throughout the twentieth and twenty-first centuries.[2] In an article titled "The Toyota Horde," military analyst William Owen offers the frightening scenario that such poorly financed fighting forces could be dramatically more successful if they just continue to embrace and increase their redundant capabilities. In particular, by spending their limited resources on many small fighting units supported by highly maneuverable small trucks (again, those implacable old Toyota pickups) with small-caliber weapons and rocket-propelled grenades, they can inflict massive damage, both military and political, on a much larger force that is typically penned into a smaller number of options for engagement. According to Owen, insurgencies don't even have to win these battles; just slowing down a heavy costly force is a small victory in itself—not unlike the natural phenomenon of "autonomy," wherein some species can shed parts of their body (like the tail of an iguana) to distract and buy time to escape from a predator.[3]

Likewise, Navy Postgraduate School professor John Arquilla warns us of "the coming swarm," referring to terrorist groups' plans to use small forces to attack multiple sites at the same time, a form of attack that will easily foil plans by well-resourced countries like the United States to use "overwhelming force" against one or two sources of attack.[4] Arquilla suggests that the only way to fight such a swarm is to create a swarm. Rather than focus on "overwhelming force" or "elite" forces, he says, we should focus on many small forces that are just "good enough." These strategies mimic nature in more ways than one—they employ redundancy *and* they exploit the pathway of adaptive evolution, which is not toward perfection, but toward competency in the current environment.

But the kind of simple redundancy that assumes and accepts the failure of many of the individual links in a security system should give us pause, because we are a species that cares for our offspring and for our fellow humans, and if you are reading this you are likely part of a society that doesn't condone mass sacrifice of people for a larger cause. The "Toyota horde" and the "coming swarm" are only effective because a lot of redundant people (in the minds of the insurgencies) are likely to die in these distributed engagements. And while insurance actuaries can work out the probabilities of death or dismemberment for people of different ages and careers to optimize the profit gained from their insurance portfolios, developing security protocols that will only work for a small subset of the population at the expense of all the rest is a non-starter, ethically and politically.

What is more feasible is to look for models from the "imaginative" redundancy built into nature. With this kind of redundancy, there are multiple problem solvers, but each employs different methods. This means that not only is there a backup if one system fails, but there is a different type of backup that won't fail in the same way against the next threat. And once again, this natural security system is found at every level of biological organization.

For example, consider how the proteins that carry out the essential life functions are constructed. An RNA (a single-strand molecule that "complements" or matches the other strand of the double-stranded DNA molecule it was created from) provides the template for each type of protein based on thousands of combinations of four base molecules (the U, C, A, and Gs of the genetic code). The four base molecules are arranged in patterns of three molecules, known as codons, to make each of the thousands of amino acids that make up a protein. There are only twenty different types of amino acids but there are sixty-four possible codon combinations of the four letters that code for them (the 4 bases in 3 possible positions $= 4^3 = 64$). From a strict efficiency standpoint, this would appear wasteful. But from a natural security perspective it makes sense. The genome is under continual attack by "translational parasites" such as viruses, which essentially mimic the RNA template of their host in order to create proteins (or in the case of HIV, double-stranded DNA) to gain access to the cellular structure of their host. The virus version of RNA, and the codons that make up the RNA, must be a good match with the host in order to get the host to effectively produce viral molecules. By having many different ways to code for its own proteins, the host can avoid getting mimicked by the virus.[5] It is a strategy essentially like regularly changing your computer password to avoid, among other problems, viruses.

What is interesting about this redundant strategy is that the basic process of RNA translation is virtually identical across most species on Earth, but there is enormous variation when it comes down to how codons are translated into other molecules like proteins. So, a remarkably simple system of replication, consisting of just four basic building blocks, provides an important line of defense to millions of organisms against millions of different viruses. It's like giving an identical (and huge) box of Legos (the old Legos, before they molded every conceivable shape for you in the factory) to millions of kids around the world and seeing the diversity of things they come up with.

This kind of more specialized redundancy is often a sign of evolutionary advancement. Sure, those multi-legged centipedes have been around a long time, but with all those legs doing essentially the same thing, they haven't been wildly successful like their cousins the beetles, which have specialized their multiple limbs into functions for flight, defense, reproduction, fooling enemies, and so forth. Beetles have come to dominate the Earth, long outlasting the dinosaurs of the Jurassic period with whom they initially evolved, and now appearing in over 350,000 species. No wonder that when biologist J.B.S. Haldane was asked by a theologian what his study of nature revealed about the Creator, he reportedly replied, "He has an inordinate fondness for beetles."[6]

Although beetles dominate the Earth, there are many more examples of this kind of imaginative redundancy in nature. We intuitively think that cactus spines are a defensive measure against herbivores, but their primary purpose is more likely to create a kind of lattice work of shade that keeps the cactus cool and reduces water loss in the brutal desert heat. And while they don't do a bad job of protection against some grazers, they actually work much better as protection for pack rats, which routinely gather viciously spiny cholla buds and use them to defensively armor their dens. This repurposing actually has a benefit for the cactus, as the pack rat spreads the cactus buds to new environments.

As a hunter, the octopus has camouflage, keen problem-solving skills, shape-shifting abilities, strong tentacles, an ink cloud to distract and hide its movements, a strong jaw, and in some species, poison. As a hunted prey, the same octopus has those same features—redundant offensive and defensive systems that all work a little differently from one another. Acacia trees have sharp thorns that protect them from grazing, but those thorns also provide a home for highly aggressive and territorial ants, which further protect the tree from grazers and parasites.

Basic repetitive redundancy is not enough to be truly successful, but even the more advanced redundancy of specialized functions would soon run into the same problems as basic repetition if those functions didn't *change* to adapt to changes in their own body and in the surrounding environments. The emergent result is more than just the increasingly popular concept of *resilience,* which implies an ability to return to (or resist being driven away from) a state of stability. The ability to change is essential because natural systems are so dynamic (this is also why some ecologists push back on the idea of resilience as being too focused on maintaining the *stability* of natural systems).

It is in fact the combination of redundancy and the inevitable variability and change of nature that provides both the drive for evolutionary innovation and the means to carry it out. In the early days of molecular genetics a scientist named Susumo Ohno postulated, mostly on circumstantial evidence, that the main force behind the creation of whole new adaptations was not natural selection, which he argued "merely modified" existing innovations, but the continual process of genome replication, which inevitably rearranged all those redundant parts through mistakes in transcription, or accidental deletions or repetitions of parts of the genome, into novel combinations.[7] While biologists agree that natural selection is still an essential force in evolution, modern molecular genetics generally supports Ohno's supposition, finding that the enormous levels of redundant features in the genome are both the building blocks and the legacy of evolutionary change.[8]

At the far opposite end of the spectrum of biological complexity, ecosystems provide a good example of how creative redundancy can lead to secure systems. Ecosystems are complex groupings of many different kinds of species (e.g., soil microbes, worms, squirrels, trees, and humans) and the chemicals (water, nitrogen, carbon) and energy (solar, kinetic, potential) that flows through the system. All those independent players and the changes and variations in each and every one of them would seem to result in a disorganized mess; but, in fact, the opposite occurs. The diversity of species present seems to impart long-term survival to an ecosystem. It can afford to lose a few species, and if the remaining species are improving their survival, they may take up the *ecological niches*—the places (e.g., deep crevices, wave-swept reef crest, holdfast of a kelp plant) or ways of living (e.g., being a predator or a parasite) left by the departed species.

Our friend Ed Ricketts, traveling with John Steinbeck to the Gulf of California, or Sea of Cortez, in 1940 recognized the ability of species to fill gaps left by removed species. The gulf is an immensely rich ecosystem formed in the long narrow gap between mainland Mexico and the Baja California peninsula, but even in 1940—before Cabo San Lucas had a single light to guide ships around the tip of the Baja peninsula (now it is a mega tourist development bathed constantly in the light of high-rise hotels, snarled traffic, and discos)—Steinbeck and Ricketts could see the emerging threats to the ecosystem. During their trip, they boarded a Japanese shrimp trawler and were horrified to see the fishing practices, which involved dragging a heavy net across the sea floor, picking up everything in its path (Ricketts, in these innocent pre–Pearl Harbor days, was also a little perplexed as to why the fishermen were taking so many depth soundings). Only a small percentage of the catch was shrimp; the rest (nowadays referred to as bycatch) was thrown overboard, usually to die because the sedentary creatures were ripped off their bottom habitat and the air-filled swim bladders of bottom fish exploded with the rapid pressure change. At the time Ricketts noted, "And it is not true that a species thus attacked comes back. The disturbed balance often gives a new species ascendancy and destroys forever the old relationship."[9]

Forty years later, in 2004, I was fortunate to take part in an expedition to retrace Ricketts and Steinbeck's famed journey, so that we could see what had changed in the Gulf of California—how the old relationships, severed by massive coastal development, intensive commercial and recreational fishing pressure, and climate change, had given ascendancy to new species filling the open niches. One of the expedition organizers was Bill Gilly, a neurobiologist from Stanford University's Hopkins Marine Station, a stone's throw from Ricketts's old lab and the Monterey fishing harbor from which

Ricketts and Steinbeck started their journey. Neurobiologists like Gilly love squid because they have the largest single giant axon nerve cell of any animal (and maybe also because the rest of their bodies are delicious when marinated and tenderized and grilled over a barbecue). That makes them a good system to study electrical impulses, ion channel function, and other things we'd like to know about nerve cell function. If regular old six-inch market squid have such lovely neurons, you can imagine what a draw the jumbo Humboldt squid, which grows to seven feet, would be to a neurobiologist. And it was these jumbo squid, reported to be amassing in huge numbers in the central Gulf of California, that really attracted Gilly to help put together the retracing of the Ricketts-Steinbeck expedition.

Curiously, although Ricketts was fascinated by jumbo squid and ordered them for his laboratory in the rare times they washed up in southern California (one of the few photos of Ricketts shows him in a rubber apron holding one of the monsters), he made no mention of them in the narrative or scientific appendix to the *Sea of Cortez* journey—he and Steinbeck just didn't see them in the gulf. Forty years later they are impossible to ignore. They amass in huge aggregations and with their strong tentacles and huge beak-like jaw eat everything, including one another. Just as Ricketts predicted, while overfishing has stripped the gulf of most of the huge top predators, like sharks and billfish, a new species, the jumbo squid, has gained ascendancy. It is the new top predator, and for now it is filling the niche left by the departed giants. But as with the old giants, the squid are attracting a growing fishing industry that is eagerly pulling in millions of tons a year of the creatures to sell to the Asian market.

The ultimate lesson from the history of the demise of the Gulf of California may mirror that of ecosystems around the world. That is, if too many species depart, the ecosystem redundancy is lost and the whole ecosystem will collapse. Biologists Paul and Anne Ehrlich likened this to taking rivets off a machine such as an airplane.[10] Pulling a few off likely won't cause catastrophic damage, but exactly which rivet, when pulled off, will lead to the wing falling off? We don't know, and it's not an experiment we want to try when our lives depend on it.

Although as a society we seem to have multiple ways to bring down even the most robust ecosystems (which is in turn beginning to impact our own survival; see Chapter 10), our own bodies employ redundant strategies for our survival quite effectively. Our senses are a good example. Having multiple ways to observe the world gives multiple lines of defense against different threats. We can smell something burning, hear someone's cries for help, see something suspicious, taste spoiled food, and feel for hidden contraband. We typically think of our sensory systems as linking a particular sense to a key part of our body that is linked to a key part of our brain—we see with our eyes, and the information is processed in our visual cortex; we smell with our nose, and the information is processed in a dense package of nerves known as the olfactory epithelium. The amount of redundancy built into these systems tells us something about how keen our senses are. Dogs, for example, have their olfactory nerves packed into an epithelium that is seventeen times bigger than that in humans.

But these relationships between senses, sensory organs, and neural processing centers are not fixed, but adaptable themselves. Geerat Vermeij illustrates this well. When he lost his sight at a young age, his visual cortex wasn't left as vacant space in his brain. Rather, he feels, as he began to rely more and more on his tactile sense, the parts of his brain devoted to tactile information processing literally took over his visual processing center, so that, in his words, he can now see with his fingers. By contrast, special education instructor Daniel Kish adapted to early blindness in a completely different way—learning early to use his auditory ability to "see" his surroundings by making clicking sounds and listening for the different echoes that came back to him when reflected off different shapes and materials.[11] In addition to hard work and prodigious intellects, environment and cooperation have been invaluable to the incredible contributions these men have made—both acknowledge the benefits of growing up with strong mentors in an age and in locations where blind people were accepted as equally valuable to society (not the case for the vast majority of blind people in human history). Yet at the core, their basic survival is attributable to having redundant *and* adaptable sensory systems.

At a higher level of human organization, redundancy is equally important. This is because a single person, despite having several different senses and a complex mind, still only observes a small part of the world. Through that limited portal, they may come up with biased or just plain inaccurate solutions to problems. Ironically, this may be especially true if that person happens to be an expert in the topical area of the problem. The "paradox of the expert," as originally described among psychologists, considered the idea that the more information your brain was storing about a particular topic, the more time it would take and errors you might commit in retrieving a particular piece of information. Although this might not be actually true due to tricks of neural architecture and practice in retrieving relevant information that experts usually possess, the term has more recently attracted a broader meaning, referring to commonly observed phenomena that sometimes the best experts on a topic come up with exactly the wrong answer.

One way to avoid this paradox is to have a lot of people, both experts and motivated nonexperts, try to solve a problem. The idea is that large numbers of problem solvers should even out the preconceived biases of experts and also buffer against the small percentage of nonexperts who come up with exceedingly bad ideas, or are maliciously trying to offer the wrong answer. Moreover, the adaptive redundancy of a lot of different minds works two ways—it can lead to a convergence on a most probable solution, or it can increase the chances of finding a truly novel solution. When it comes to security, we can't ethically conduct experiments on the effectiveness of redundant problem solvers. Fortunately, there are already both well-established and newly emerging forays into this form of problem solving.

USING REDUNDANCY AS A NATURAL EXPERIMENT

When, as a boy, I would accompany my grandfather on his trips to the Aqueduct or Belmont race tracks, my bets based on the most cleverly named horse rarely won, but they also didn't significantly skew the odds relative to the contributions of thousands of presumably better handicappers than I. It is the collective wisdom of those handicappers, translated first through dollars bet and then into the statistical measure of the odds of a horse winning, that turn out to be fairly good predictors of each horse's success.[12] The idea of pari-mutuel betting itself is based on the principle that many redundant actors will trend toward the correct solutions, making up for the small number of individuals who do very stupid things (like bet on the horses with the best names).

After the financial collapse of 2008, gambling and financial markets seem to be forever entwined, so it's appropriate to recognize that the same logic of pari-mutuel betting underlies futures markets, which are financial instruments that allow investment based on the expected changes in the prices of commodities and other traded goods. It has been pointed out that futures markets have been used at least since the time of Aristotle,[13] but their origins in nature go far deeper. Ants and other social insects, such as honeybees, send out individual scouts to search for better, safer homes, but they won't move as a colony until a threshold quorum number of scouts indicate the attractiveness of the new site.[14]

These types of futures prediction schemes have been subsumed lately under the term crowdsourcing,[15] which has been applied to energizing wide swathes of the public to improve business practices, guess the outcome of elections, develop translation services for obscure languages,[16] engage in social movements, and determine how philanthropic resources should be distributed.

My friend Josh Donlan, who has a knack for proposing clever and controversial ideas that cross the borders between scientific ecology and public policy (he recently proposed "rewilding" North America by bringing back descendants of the massive mammals that once roamed the land, including camels, cheetahs, and lions[17]), has proposed a highly controversial futures market for endangered species[18] in which the government would sell futures on species at risk—if the species population or critical habitat area decreases, investments accrue to conservation funds; if the species populations improve, investors receive a return on their speculation.

The notion of betting on the future rubs many people the wrong way for many reasons. A recent proposal to allow people to invest in movies they thought would become blockbusters or not was vehemently opposed by the film industry, which feared people would manipulate the system and cause financial ruin of expensive movies,[19] and was also opposed by film lovers on philosophical grounds: they felt it would turn their art into just another commodity.[20] Some critics think that certain things shouldn't be traded but heavily regulated by the government. For example, we shouldn't have to speculate on the future of endangered species because the government should just do its job of protecting them. This is similar to criticism raised against "cap and trade" programs to reduce pollution by allowing heavy polluters to buy pollution credits and efficient producers to sell credits, with the total number of credits capped at a certain amount. Opponents to these approaches feel that selling rights to pollute, or providing a way to make money off failure (a bomb at the box office, a woodpecker going extinct) is dismally cynical or even unethical.

Pentagon officials found this out the hard way when DARPA (the same agency that created the Grand Challenges) floated the idea for a "terrorism futures market,"[21] which would have created a website for people to place bids on when they thought the next terrorist act or assassination would occur. Senator Barbara Boxer called the idea "very sick" when she raised the proposed project by surprise, just days before it was supposed to go live, during an unrelated hearing in the Foreign Relations Committee. A chorus of angry lawmakers from both parties quickly seconded Senator Boxer and called for firing the DARPA spooks who thought up the idea. The program was terminated almost immediately.

That many economists and security analysts still think a terrorism futures market is a really good idea[22] is a clear illustration of a point I made earlier: even biologically inspired or biologically analogous solutions must pass through ethical, political, economic, and other social filters before they can be made practicable for society. This does not invalidate them—in fact, all solutions in nature go through many filters before they become adaptations that stay with organisms through their life and across generations. These filters may be the presence or absence of predators in the region, the particular environmental conditions of the time and place, the disease load carried by the organism, and, in more social organisms, very similar filters to those faced by human societies—an efficient new gathering strategy devised by a low-ranking chimpanzee, for example, might not get replicated just because of her status in society. The point is that these evolutionary filters are another example of redundancy that both ensures the elimination of bad ideas and strengthens those that survive. Rather than dismiss the various ethical, moral, economic, and political objections that are often raised when considering a change in policy, they would be better embraced and used as tools to strengthen the adaptation.

Of course, the best example of an adaptive redundant system in society is the Internet. There are now billions of redundant observers and disseminators of information operating in a network loosely draped on a minimal architecture that is itself continually changing. Along with the rise of the Internet came a whole new field of science studying the properties of networks. The more scientists looked at human-made networks like the Internet, the more they saw nature. Network analyst Alessandro Vespignani has noted that the Internet "has become one of the first human artefacts that we study as a natural phenomenon."[23] The Internet is not that different from a network of terrorists plotting an attack, and both of these types of networks are not that different from organisms in an ecosystem or the relationships between different proteins in a yeast cell. There are striking similarities in how they form, how they respond to change, and how they are vulnerable.

Networks emerge in similar ways, through preferential links made between independent operators. In network science, these operators are called nodes. Your personal website could be a node, and maybe it's linked to the company you work for, your friend's blog, some websites you particularly like, and a search engine—most likely Google—but not the website of someone you've never met or a company you've never heard of, even though *they* are likely linked to Google as well. As a result of this preferential linking that doesn't treat every single node equally, a small number of nodes grow to be incredibly important hubs of activity, and many nodes remain connected to just a few other places on the net. Networks in nature don't build much differently. Hermit crabs in a network of tide-pool organisms may be linked to snails because they use their shells, and both species might be linked to a larger crab species that eats them. A generalist predator, which eats a lot of different kinds of organisms, will be a hub in this network. So too, could something like the tiny algae that coats the rocks and is grazed by many different species.

Understanding the roles and relationships of entities in a network, which is essentially what all ecologists do, has value far beyond developing plot lines for *Animal Planet* documentaries. Eric Berlow, an ecologist who studies food webs, argues that network science is primarily valuable because it can turn a *complicated* problem into a much more solvable *complex* problem. In a brilliant three-minute TED (technology, entertainment, and design) talk in Oxford in 2010, he used the same network analysis he uses on Sierra Nevada lake food webs to transform an über-complicated government-created diagram of U.S. strategy in Afghanistan—a tangled yarn ball of lines and arrows and boxes that was howlingly ridiculed in the media—into a very small set of truly actionable tasks.[24] Instead of focusing on the whole complicated picture at once, Berlow highlighted the networked *relationships* between entities in the U.S. strategy and then cut out all the entities that were more than three or four degrees removed from the ultimate goal of "increasing popular support of the Afghan government." Of this much smaller remaining set, he also eliminated the entities that no one could do anything about, such as the harsh terrain of Afghanistan. In less than three minutes, his analysis collapsed the entire complicated affair into just two necessary actions: active engagement of ethnic rivalries and religious beliefs, and fair transparent economic development. Both complex tasks, to be sure, but much more clearly understandable than the tangle of yarn the United States has been tangled up with in Afghanistan for over a decade now.

Like natural systems, networks respond to change quite rapidly. Albert-László Barabasi, whose book *Linked* is the most accessible and entertaining treatment of network science, likens networks to our own skin.[25] Consider how much your skin can change—the nerve cells on your arm are just a few of millions, little mildly important nodes working away as part of a vast network. If you spill hot bacon grease on your arm, suddenly these nerve cells become the most important hub of activity—screaming out directions to the entire body and getting responses from the entire body, "Step away from the oven! Get the grease off! Send in the pain killers! Send in the immune system to fight off infection!" Likewise, consider how many sites that were once just lowly little nodes—someone's personal project—found a niche and blossomed into the next go-to source for instant celebrity gossip, extreme political hyperventilating, or online shopping.

The Internet, like natural systems, doesn't have any end goal or overarching values. So if change steers it toward providing the ultimate platform for global meet-ups of wannabe radical jihadists, that's where it will go. Indeed, the Internet has evolved quite well with the changing face of global terrorism. It was once an outright recruitment tool for drawing fighters into al-Qaeda and other terrorist groups. But as large terrorist organization became increasingly forced underground after 9/11, and as intelligence services scrutiny of jihadists websites grew, the recruitment mission largely dropped off. In its place, the sites of radicals became places to amplify the alleged insults of the foreign occupation of Iraq (a more effective, though indirect, means of recruitment) by posting the latest videos of allied atrocities such as the torture of prisoners at Abu Ghraib or the recorded killings of Americans by IED attack. More recently, chat rooms stemming from radical sites have been the places where disaffected youth from all over the world have been meeting up to flex their ideological street "cred" and share dreams of martyrdom.[26]

Can networks of would-be terrorists on the Internet, or the real-world networks of terrorists that arise from them (or more likely, from a number of kinship relations between radicalized individuals), be taken down? Like ecosystems, all networks are resilient to attack, to a point. If an attack takes out a node, like your personal website, the network will carry on just fine—this is akin to an individual animal dying in an ecosystem. It has been noted that by the time of the 9/11 attacks, several of the key hijackers on different planes had no close networked connections between one another, so that even if some had been caught, the rest of the terror network would have likely survived to carry out the rest of the attack.[27] Even if an attack takes out a pretty important hub (akin to losing a whole species in an ecosystem), there may be some inconvenience, but as with the ecosystem "a new species will gain ascendancy," and the network will anneal itself around the gap left by the departed hub. But as the threat gets larger, either by winding its way through many nodes in the network or by taking out some really big hubs, the resilience of the network is much less certain.

We are learning now that with strong interconnection, we can also see catastrophic failure. Ferenc Jordan is a Hungarian scientist who studies all sorts of networks, from children's social groups in schools to the power dynamics of social wasps to food web networks that bind different species in an ecosystem. He is especially interested in how these networks change when a big event happens—a popular child leaves the school, the queen wasp dies, or uncontrolled fishing removes a top predator. When a much bigger human tragedy occurred on July 7, 2005, Jordan was ready with the tools he uses to study networks to examine the attack on four London underground stations. He found that of 3.2 million possible combinations of stations to bomb to achieve maximum structural destructiveness to the underground network, the terrorists chose the second most destructive combination.[28] Whether the terrorists actually used network analysis, just chose the most intuitive or easiest set of stations to attack, or just got lucky, is unclear, but it suggests that studying networks has enormous value both offensively and defensively.

As our world becomes more interconnected, networks can actually become more vulnerable. This is because most networks now are not so isolated, but rather have become networks of networks. Recent analysis by Israeli and American scientists published in the journal *Nature* has shown that when networks are *inter*dependent, failures in the nodes of one network will likely lead to failures of the network it is connected to.[29] They also found that more heterogeneous networks, meaning those with fewer redundant copies of individual nodes, are more likely to collapse.[30]

The same day that the authors' rather theoretical paper appeared in *Nature,* it was validated in the real world through the misery of thousands of travelers stranded throughout Europe. A single volcanic eruption in Iceland, combined with just the right wind conditions, grounded all flights from most of Europe for several subsequent days. As flights were cancelled and passengers scrambled for other ways to get to and out of Europe, the entire transportation system came to a standstill. No planes, no trains, no boats, no rental cars. People were forced to camp out in airports, ironically stuck at an absolute standstill while located in some of the largest hubs of one of the largest transportation networks in the world.

Why didn't redundancy work in this example? There are a number of likely reasons. For one thing, the threat was completely novel and was huge, encompassing the entire airspace very quickly. The airlines had little experience figuring out new flight patterns, safety precautions, and maintenance procedures for a giant ash cloud. As they began to compile this information in the days following the eruption, conflicting reports emerged, with some commercial airlines reporting little problems in test flights and the U.S. military reporting some worrisome conditions and deterioration of equipment during its test flights. One big contributor to the problem, seized on by critics of a single European Union, was that the EU treated all of the transportation network as a single entity, equally disabled by the ash. Only after many days of misery did the EU concede to divide its airspace into three units (a marginal improvement over just one unit), with different abilities to allow flights depending on local conditions.

Interconnected redundant systems solve plenty of security problems, but they also create literally a world of new security problems. This is because as ever more clever problem solvers interact, they elicit responses among their adversaries, who themselves must improve in order to keep up. The resulting escalation of armaments and defenses, and strategies and counter-strategies, that occurs between competing organisms (which is the topic of the next chapter) is both a response to the creative ways organisms use redundancy and a force for further adaptation. Just as some organisms use redundancy better than others (think beetles vs. centipedes), some organisms play this game of escalation better than others. The Cold War gave us an artificial sense that escalation was mostly about who had more, or more powerful, weapons, but as the next chapter shows, this kind of linear escalation is fairly limited in nature, and since the current state of security is far more wild than it was during the Cold War, we might want to pay attention to how biological escalation really works.

chapter six

BEYOND MAD FIDDLER CRABS

In the 1983 film *War Games,* two teenagers inadvertently nearly set off World War III when they hack into the computer of the North American Aerospace Defense Command (NORAD) and cause it to simulate a full-scale nuclear attack from the Soviet Union. In desperation they track down the disenchanted computer scientist, Stephen Falken, who programmed the computer, and hightail it in his helicopter to NORAD to stop the U.S. Air Force nuclear hawks before they unleash their missiles for real. As they watch the out-of-control computer, Dr. Falken shares a plan with the movie's female lead:

> Falken: Did you ever play tic-tac-toe?
>
> Jennifer: Yeah, of course.
>
> Falken: But you don't anymore.
>
> Jennifer: No.
>
> Falken: Why?
>
> Jennifer: Because it's a boring game. It's always a tie.
>
> Falken: Exactly. There's no way to win.[1]

Accordingly, with the air force brass nervously watching on, they program the computer first to play tic-tac-toe against itself until it gets bored. Then they have it play thermo-nuclear war games against itself until it likewise figures out that there is no way to win and it finally stops the real game it is playing with the U.S. nuclear stockpile.

It's a captivating scene because the simplicity of the tic-tac-toe board and the horrifying trajectories of thousands of nuclear missiles painting a web around the world map contrasted on the huge NORAD video display exemplify the same bare logic—some conflicts can't be won.

At the end of *War Games,* the audience breathes a collective sigh of relief. The DEFCON level is returned to low, stability has returned. The United States and Soviets both still have thousands of missiles pointed at one another, but the computer running them has learned, just as the humans who made the missiles and the computer programs, that using them would benefit neither side. In the fantasy world of *War Games* and in the real world of the Cold War, this stability was known as "mutually assured destruction," or MAD. The idea that any release from this stable state would lead to immediate and complete catastrophe became the mechanism and the energy to maintain stability. MAD is what kept the Cold War cold.

As a child and budding marine biologist, I was also fascinated with mutually assured destruction. But I observed it in the salt marshes of Cape Cod Bay, where I could watch fiddler crabs for hours. These are small crabs that live in large colonies in burrows on the edge of marshes. The males are distinctive in that they grow one enormous claw, which they wave menacingly at other males in displays. Two males competing for a female will wave their large claw at one another until one of them backs down and the other claims the spoils of victory. The crabs apparently have an incredible innate capacity to size up one another's claws, because the one that backs down is almost invariably the one with the smaller claw. But oddly enough, they rarely, if ever, fight with these huge weapons of destruction. It's as if they implicitly understand that full-scale claw-to-claw combat would leave them both battered and unable to feed or mate.

Fiddler crab security works in the following situations: (1) where both sides of the conflict have similar resources (neither crab has the ability to suddenly grow three times as big as the other), (2) where both sides of the conflict adapt in similar ways (neither crab is able to suddenly grow wings and attack the other from the sky), and (3) where both parties share a common end goal (both crabs want to win a female to mate with).

In other words, this kind of stability-inducing conflict has nothing in common with today's security problems. Almost all of today's security problems deal with unequal resources (e.g., al-Qaeda vs. the U.S. Department of Defense; failing subsistence farms in Kenya vs. massively subsidized U.S. corn), unequal pathways of adaptation (insurgents changing practices on the fly vs. a U.S. Army forced to follow "SOPs"), and different end goals (hackers trying to cause cyber chaos vs. you trying to finish your report by the close of business).

All this inequality and disparity of goals creates imbalance and instability, when what we'd ideally like in any security situation is *stability*. Fiddler crabs and MAD give us a nice model of stability, but it's a fairly narrow one pertinent to a particular set of historic and evolutionary contingencies. Though I still have an inordinate fondness for fiddler crabs, I've had to move beyond them in my biological studies in order to begin to understand biological complexity. The Cold War, like fiddler crabs to a budding marine biologist, is a fascinating focus of conflict studies and history, but it's hardly representative of the current state of complexity.

Are there more general models of stability in nature? We might immediately think so, having been inundated throughout history with soothing words from poets, writers, politicians, and environmentalists of "harmonious nature" and the "balance of nature." From a scientific standpoint, my colleagues in ecology have been looking for the existence of "stable states" for decades. Do they exist? If you squint enough—that is, look from far enough away, don't get into the details too much, and paint in broad strokes—you can see stable states in ecology. But most ecologists now acknowledge that stable states are a good abstract generalization that helps us think about changes of various levels of severity, but stability can't possibly describe real ecological systems at the fine scale.[2]

The tide pools I study at Hopkins Marine Station in Monterey are still much more like they were when I first saw them in 1993 than like they were when they were first studied by Willis Hewatt in 1930. The overwhelming signal of change between the early study and our later studies seems to be related to warming climate. There might be a temptation to say that the tide-pool community, after a period of struggling with warmer temperatures, has reached some climate-changed steady state. But I know from my frequent visits that even within the realm of a tide-pool community more dominated by warm-water animals, there are frequent changes in who is dominant and who is on the verge of disappearing, who has taken over the most space on the rocks, and who saw a brief period of flourishing populations but is now on the wane. I'd be very nervous to use the word *stability* around my tide pools or any other ecological system.

True, there are general patterns that can be used to describe certain ecological landscapes. I can say "oak-chaparral woodland," and any ecologist and many Californians will have a fairly similar picture in their head of dry grasses, grayish-green shrubs, and gnarled oak trees scattered up a hillside. But look a little closer and some of those woodlands will be guarded in various places by squawking red-winged blackbirds perched on invasive fennel stalks, and others will be on the verge of collapse as a winemaker's bulldozer seeks to expand its winery's vines yet further, and others will be blighted by sudden oak death.[3] Ecologists may say I'm cheating by invoking bulldozers and other impacts of humans on a pure ecological concept, but the impacts of humans have become inseparable from nearly every ecosystem on Earth, so if ecological landscapes are not stable in the face of human intervention, they're just not stable. This is the world we live in.

So, what about human-created ecosystems? If we can create ecological instability everywhere on Earth, is there anywhere we can create stability? That is, can deliberate attempts to merge nature and technology create stable, self-sustaining biomes? I've seen small glass globes with some water, algae, and maybe an invertebrate like a small shrimp sold (usually in the *SkyMall* catalog) as self-sustaining ecosystems. But even those get an influx of sunlight, and on the scale of sustaining interest they fall somewhere between tic-tac-toe and a goldfish. In other words, they lack the complexity of real ecosystems.

A much more interesting experiment in stability was conducted in the desert not far from my home in Tucson. There, Texas oil heir and environmentalist Edward Bass financed the creation of a massive glass greenhouse that was to house eight humans and everything they needed to live for a two-year period. The Biosphere 2 (named as a sequel to Biosphere 1, planet Earth) mission drew huge publicity and ultimately some notoriety, as rumors of tribal infighting amid crop failure and malnutrition among the "Biospherians" surfaced. A number of technical issues emerged, like excesses of carbon dioxide caused by out-gassing from the concrete foundation. From a biological standpoint, ecological interactions didn't play out as they did in the much larger landscape of planet Earth. Pollinators died out, and cockroaches took over.[4]

Biosphere 2 is now managed by the University of Arizona as an experimental facility. A visit to Biosphere 2 leaves one gasping at the audacity of its original plan in the face of the overwhelming complexity of recreating the Earth's living biosphere. Simultaneously, you get both a tantalizing feeling of how close it might have come to achieving its dream—the plant life appears as lush and abundant as the Garden of Eden—and many reminders (like condensation dripping down every wall and electrical junction box of the musty subterranean mechanical infrastructure) of just how much farther they had to go. It remains an amazing structure for experimentally exploring things like plant growth under a climate-warmed world, but no one still harbors any illusions that it can function as a closed, stable self-sustaining ecosystem.

Thus, even ecosystems that are persistent enough to be named ("rocky intertidal," "oak chaparral") are not that stable, and the chimeras of human technology and ecology are even less so. Have I drawn too strict a boundary around the concept of stability? Any decision to define the rules of stability will be fairly arbitrary, but just for argument, let's say that a system that has stuck around in its present condition for millions of years makes a pretty good case for itself that it is indeed relatively stable. Geerat Vermeij cites the example of deep-sea brachiopods—ancient invertebrates that sort of resemble a clam on a stalk. They haven't changed much from their fossilized ancestors. They live harmoniously in the quiet deep sea, but like the sealed glass "ecospheres" sold in the *SkyMall*, they're not all that interesting. Or, to use Vermeij's more precise analogy, they lack economic power. That is, they don't command a lot of biological resources; they don't expand out of their narrow niche, they don't cause changes in other species, and they don't move mountains.

Could we find blissful stability in society? In a world of food insecurity, cyberattacks, terrorism, and terrifying natural disasters, there is something appealing about the simple life of a deep-sea brachiopod, unchanged for millions of years in its silent world where it gets everything it needs and nothing more. I imagine, for the right price, there are little goat cheese farms in southern France or Belgium that you might buy and achieve something of this existence. You could live like the little shrimp in the ecosphere, more or less getting your daily needs met, maybe trading some goat cheese for a baguette and some marmalade. It sounds lovely. For a week or two. Until you got a craving for cow cheese, or, God forbid, a Big Mac. And even life in the glass sphere has to change. I met a man in far upstate New York who bought a grove of sugar maples and hoped to get away from it all through a life of making maple syrup. Unfortunately, climate warming has invaded his stable retreat, forcing him to continually adjust his sapping dates, which once were stable enough to be fixed in north woods lore.

Unless we'd like to live in a country or a little village that never learns, never grows, never trades with outsiders, and never brings new people in or sends people out, we are unlikely to learn much from the rare stable states in nature.

BEYOND STABILITY-ESCALATION

Stability in international affairs, even the tenuous trigger-point stability of the Cold War, is likewise rare and exceptional. Even for much of the Cold War we didn't necessarily think a nuclear conflict was unwinnable. In *War Games,* the computer learned about MAD in a couple of tension-filled minutes in the multiplex. In reality, the superpowers learned about MAD and ultimately came to accept it much more slowly. In the 1960s John F. Kennedy effectively turned the ominous concept of a "missile gap"—the number of fewer nuclear missiles the United States supposedly had relative to the Soviets—into a campaign weapon. Civil defense public service announcements insinuated that it was a patriotic duty to learn to survive the initial impact of a nuclear strike so as to be able to rebuild American democracy afterward. Nuclear survival drills were routinely part of my elementary education up until the late 1970s.

It was essentially a series of ecological studies that put an end to nuclear war survival scenarios and kept the concept of MAD pertinent until the end of the Cold War. First, in 1980, a clever interpretation of many layers of geological, climatological, and paleontological data assembled by a father-and-son team—Nobel Prize–winning physicist Luis Alvarez and his geologist son Walter—revealed a wholly plausible theory to explain the mystery of why the dinosaurs disappeared at the boundary of the Cretaceous and Paleogene periods 65 million years ago.[5] The Alvarezes contended that a large impact from an extraterrestrial object could have created huge firestorms and dust clouds so dense as to block significant solar radiation for an extended period of time, killing off photosynthetic plant life and especially large organisms dependent on this life. The evidence at the time was circumstantial—one major clue was a geologic layer of iridium found in many sites throughout the world corresponding to the geologic age of the extinction event. Iridium is extremely rare at the surface of the Earth. Its most likely source in such quantity to be available throughout the world at the same time is an extraterrestrial asteroid, some of which have been found to be enriched in iridium. Although the Alvarez team and their colleagues could not at the time provide the smoking gun of a crater large enough to indicate such an impact, such a crater was discovered ten years later using advanced technology, hiding a kilometer underground in the Yucatan Peninsula region of Mexico.[6]

Because everyone is captivated by dinosaurs, or at least knows a young boy or girl who is, the unraveling of their mysterious disappearance was one of the hottest science stories of the time. The idea that forces of nature, not just some tooth-and-claw combat to the death, could fell all those mighty giants and fundamentally change the face of the Earth was both captivating and terrifying. This new view of nature's dynamics acting on a global scale essentially greased the wheels for widespread acceptance of the subsequent "nuclear winter" theory, which would outline the extinction not of our long-passed reptilian ancestors, but of humans themselves. Presented in a 1983 paper by a team of well-established physicists including the widely known Carl Sagan,[7] the theory suggested that the simultaneous explosion of multiple nuclear warheads, as could be expected in a conflict between the United States and the USSR, would lead to massive firestorms and a significant decrease in incoming solar radiation due to dust in the atmosphere. The resulting global cooling would kill off virtually all life on Earth. The idea that such a global extinction was possible was already well-primed in the public imagination. The dinosaur story even posited roughly the same mechanism; the only difference was the delivery vehicle—an intercontinental ballistic missile rather than an asteroid.

As people will do, there were different reactions to the nuclear winter scenario. All considered it an appalling and unacceptable end to the Cold War, but, as expected, there were different reactions to the nuclear winter scenario. Advocates of disarmament used it to bolster their cause, arguing that the consequences of even a small number of nuclear strikes were too terrible to allow for any policy but complete drawdown of the missiles. More hawkish types, such as President Ronald Reagan, used this frightening outcome to justify a seemingly simple solution—a missile defense shield above the Earth that would destroy nuclear weapons out in the harmless vacuum of space, before they could cause the type of global destruction that the nuclear winter scenario foretold. It was a vision of pure science fiction, echoed in the nickname it quickly acquired—Star Wars—after the immensely popular movies of the time. But its reality was science fiction, too, since long before and long after Reagan's time.

In the Smithsonian Air and Space Museum in Washington, D.C., there is inadvertently some record of how long the dream of a true missile defense shield has gone unrealized. There, next to a replica of the *Explorer I* satellite, on a framed January 31, 1958, newspaper front page displayed to commemorate the U.S. entry into the space race, is a small article set off to the side and below the fold commenting that the United States was to begin development of a missile defense system. Thus began two races of escalation with very different outcomes. Just over ten years after the rocket launch that put *Explorer I* into orbit, the United States proudly sent men to the moon, beating out the Russians, who up until that point had preceded the United States in nearly every major step toward that goal. By contrast, *fifty* years after embarking on a missile defense plan, the United States still has nothing close to a workable missile defense system.

I tend to be a technological optimist—I'd like to think that given enough time and money, some of the most difficult technological problems can be solved. In that sense, I'm sympathetic to the missile defense supporters who argue that past failures and slow progress are no reason that missile defense can't work. My concerns about missile defense are much more organic. First, all organisms must balance resources—the creature that spends too much energy on defense won't have enough left to forage for more food or to mate. Although at times it seems like we have infinite resources for defense projects, we don't, and missile defense has been sucking up a large portion of those resources for some time now. But the second organic argument against missile defense, more germane to this chapter, is that missile defense creates a dangerous escalation where you once had a rare instance of stability.

Missile defense creates escalation through several pathways, none of which are good for security. It creates an incentive to produce more missiles, under the strategy that even a working missile defense system can only shoot down some fraction of the incoming missiles, not all of them, so you might overcome such a defense through sheer numbers. Of course, few nuclear-capable regimes have the capacity to produce many nuclear-armed missiles. For these regimes, which are typically the "rogue" nations and entities like North Korea and Iran that we are most concerned about, two other escalatory options remain. They can produce more deadly weapons, such as missiles that disperse their nuclear payload into multiple warheads that spread destructive impacts, or at least lethal radiation, over a larger area. Alternatively, and more cheaply, they can choose to deliver the nuclear payload by something other than a missile, such as in a bomb smuggled inside our borders in a shipping container. These adaptations *may* happen without nuclear defense, but a nuclear defense system virtually ensures that they will happen. Missile defense essentially serves as an enzyme to speed up the escalatory process.

Arms control advocates early on recognized the escalatory power of building up defenses. That is one of the reasons why both early nuclear reduction treaties and proposed treaties such as the "New START" treaty between the United States and Russia specifically outlaw defensive structures such as hardened missile silos or defensive weapons launched from offensive silos or submarines.[8] Nonetheless, the apparent need for missile defense has become so dogmatic in the United States that politicians and pundits from the left, right, and center consider any attempts to curtail the development of missile defense non-starters for future nonproliferation treaties.

LIVING WITH ESCALATION

Accordingly, with us hell-bent on creating escalation even out of stability, and with few natural models of stability to guide us anyway, we need to understand how to survive and thrive in a constantly escalating world. We need to look where conflict still brews, where armaments and defenses get yet more deadly and more effective, and where strategies get ever cleverer; where the tiniest organisms can wreak havoc on the most powerful leviathans, where senescent creatures suddenly awake and throw their once-placid surroundings into turmoil. Fortunately, this describes just about all of nature, so we have a lot of material to choose from, but where do we start looking?

We could start by staring deep into the looking glass. That is, we can take a page from Lewis Carroll's fantasy *Through the Looking Glass,* where we find the Red Queen warning Alice, "It takes all the running you can do, to keep in the same place." Biologists have seized on this comment as convenient shorthand for the daily adaptational struggles of organisms. The "Red Queen hypothesis" states that in order to maintain the same level of fitness, or ability to survive and reproduce, relative to one's competitors and antagonists, an organism must keep adapting. The logical outcome of Red Queen dynamics is an evolutionary arms race in which opposing sides continually increase speeds, armaments, defenses, and tactics, sometimes exploiting whole new resources, in order to be able to survive and thrive.

Surely this type of escalation cannot sustain itself forever? As it turns out, 3.5 billion years of experience with life on Earth tells us that it can. Billions of years ago, very few ecological niches were exploited. That is, most of the places organisms could live were still empty. The type of energy systems now used by organisms were almost completely untapped. There were abundant openings for the jobs of predators, prey, carrion feeders, and parasites.

Fast swimming, flying, running, and swinging from trees that hadn't yet evolved to exploit untapped areas for photosynthesis were inventions still billions of years off. But all of these places and ways of living—ecological niches—would ultimately become filled. Much of this niche filling happened during brief explosive periods of diversification, especially around 550 million years ago, when most of the basic body plans of animals emerged. Then again, about 489 million years ago many new species within those basic body plans emerged, and they stretched out from the sea floor and began to dig into the sediments and explore the water above.[9] This increasing exploitation of different niches throughout evolutionary time is a really good indication of escalation between organisms. If there was no reason to grow higher to compete for light or avoid herbivory, we wouldn't have trees. If there was not a vicious battle for resources in the seas and abundant unexploited plant life on land, we probably wouldn't have gotten amphibians. It is true that sometimes, outside forces like asteroids and ice ages wreaked changes far more powerful than the day-to-day arms races, creating mass-extinction events in several cases. Yet after every one of these events, life, and its attendant escalation, rocketed back.

Because much of this evidence of escalation is indirect, Geerat Vermeij sought to take one thick slice of geologic history—the past 550 million years or so—and test three specific hypotheses, using actual animal fossils, to see if arms races and escalation did occur.[10] Vermeij looked at whether traits relevant to antipredatory or competitive ability, like thick shells or strong claws, became more common through time. He then looked at the "adaptive gap" between organisms' capabilities and their environments to see if, through time, organisms have become better adapted to their environment. Finally, he looked to see whether the biological hazards of competition and predation have become more severe over time. Vermeij's findings from the rich store of fossils he studied confirmed what general trends suggested for the entire history of life—it gets harder and harder to survive on Earth, but this struggle drives unparalleled innovation and unimaginable diversity.

Consider how this plays out in nature. Sea anemones were among the first multicellular life forms on Earth, not that far removed from sponges. They have a pretty simple way of going about their business. They are basically a voracious mouth surrounded by a ring of tentacles. They are supported and attached to the rocks by a soft fleshy stalk. They stay in one place waiting for food drifting in the water to flow by their mouth or get stuck in their outstretched tentacles. But as coastal seas became more crowded with predators and even other anemones, their soft body and sessile nature made them fairly vulnerable. So through time they developed stinging cells in their tentacles, which, like the best of adaptations, serve a multitude of functions. They can stun or kill live prey, they protect the anemone from predators, and they even ward off competing anemones who might try to take their prime real estate on the rocks. With certain clonal anemones, you can see a bare rock "demilitarized zone," like the line between North and South Korea, between two growing colonies that use stinging cells to repel one another when they meet. In addition to developing stinging cells, some of the anemone line split off and became free-living jellyfish—thus exploiting a whole new niche of the open ocean.

These innovations in turn helped to drive innovations among completely unrelated species. Out at sea, the homely *Mola mola* (ocean sunfish) evolved to exploit the niche created by teeming numbers of jellyfish in the open ocean. Back on the coastal rocks, where the struggle for life is perhaps as fierce as anywhere on Earth, biological escalation, some of it in response to the armament of anemones, drove some remarkable adaptations. Under the extreme predatory pressures of the rocky shore, even the stinging cells of the anemone became a whole new niche to exploit. Specifically, nudibranchs, or sea slugs, developed the ability to ingest anemone stinging cells whole and use them in their own defense.

In human tide pools—those environments of extreme escalation, such as a theatre of war, humans also expropriate their enemies' armaments for their own ends. Much of the explosive materials now used in IEDs that are deployed against U.S. troops in Iraq and Afghanistan can be traced to weapons caches left by the Soviets after their long and unsuccessful occupation of Afghanistan. In general, insurgents are a particularly vexing problem for regular armies to fight, because they seem to play the escalation game so well. Ground observations by counterinsurgency officers show that the average time for insurgent fighters to adapt to new tactics, techniques, or procedures of U.S. troops is about fourteen days.[11] U.S. Secretary of Defense Robert Gates remarked with frustration at a congressional hearing in March 2007 that "as soon as we . . . find one way of trying to thwart their efforts, [the insurgents] find a technology or a new way of going about their business."[12]

Observations like these have prompted speculation that insurgencies can adapt faster than regular armies, such that in essence we will always be in a losing situation. This immediate reaction ignores the fact that at the ground level, soldiers and platoons in regular armies adapt pretty well themselves. Naval Postgraduate School counterinsurgency expert James Russell says it's "nonsense" to reflexively describe insurgencies as more creative, flexible, or adaptable.[13] Moreover, speed of adaptation is not the same thing as effective adaptation. In nature, adaptation happens at a huge range of speeds. It is certainly not only the gradual accumulation of change across millions of years of evolution, as many believe.

Detailed studies of Darwin's finches in the Galapagos have shown that natural selection acts rapidly on each finch species' adaptations for different types of food sources.[14] Flowering plants adapt toward self-pollination within a few generations when they lose their pollinating insects.[15]

The point is that what matters is not the speed of adaptation, but what problems it helps you solve and what problems arise as a result of an enemy's adaptations. As Michael Kenney found in analyzing escalation between drug-trafficking "narcos" and the law enforcement "narcs" who try to stop them, "[trafficking networks] enjoy a number of advantages over their state competitors—among them stronger incentives to adapt, smaller coordination costs, flatter organizational structures, and fewer institutional impediments to action. These advantages influence, but do not determine, outcomes in competitive adaptation."[16] Colleagues of mine are currently poring through data on insurgencies to really pin down whether they adapt faster or not, but what is relevant is that they adapt fast enough to require us to adapt ourselves, and they adapt *differently*. It would at least be useful to understand why they adapt differently.

Indeed, variation and difference between two entities involved in an escalating evolution is always the key to understanding what drives the escalation. Whereas the rare cases of evolutionary stability arise from the similarity between organisms (the fiddler crabs claws are so close in size), evolutionary escalation of conflict occurs because of the differences between them. I'll focus here on three important sources of differences between escalating organisms: the resources available to either side, the motivation to change in either side, and the way in which each side uses information and communicates.

Difference in resources seems like an obvious source for adaptation and change. Shouldn't the organism with more resources at its disposal be able to evolve more rapidly? At least, shouldn't the better-resourced side feel more secure? It would seem that if we have a lot of money, we can buy a lot of different things to ensure our survival—fire extinguishers and burglar alarms and bullet-proof glass and expensive Swedish cars with five-star safety ratings and maybe a few bodyguards as well. The Powell Doctrine, which guided the U.S. military's initial foray into Iraq, buys into this logic. The doctrine stated that the United States should not enter a conflict unless it has overwhelmingly superior resources to apply to the job. In part this doctrine was engendered by interpretations of the failures of the Vietnam War that asserted that the United States simply did not apply enough resources to the battle. But just as Powell's assertion before the United Nations that Iraq was developing weapons of mass destruction turned out to be a poor reflection of the situation on the ground, the Powell Doctrine was poorly matched to the way the war escalated in Iraq. The utility of an overwhelming force doctrine was justified in the lead-up to the war by (it turns out later) hyper-inflated assessments of the abilities of Saddam Hussein's elite conventional forces. In reality, the fearsome Republican Guard crumbled easily under moderate pressure—no overwhelming force was needed—but the defeat of the conventional forces opened a new chapter of insurgency, for which superior resources were of little value.

Even a cursory glance at nature tells us that it is foolish to think that resources are enough to come out ahead in evolutionary escalations. Viruses mutate with very few resources yet cause huge security issues for vastly better resourced humans. And small individual organisms that each use few resources, like locusts, can wreak havoc when their actions are matched by millions of other individuals. We have poured billions of dollars and countless hours of research effort into developing herbicides, pesticides, and antibacterial agents, yet the results are numbingly similar every time. The weeds, the bugs, and the microbes, each with very few resources, come back stronger, more destructive, and more deadly than ever. A quick glance at the news shows us the results of these escalations in which the far less resourced side is decidedly winning. Farmers, after years of applying the supposedly harmless herbicide Roundup, are now resorting to ever more toxic chemicals and more destructive tilling practices to deal with outbreaks of Roundup-resistant "superweeds."[17] Bedbugs, once the scourge of city dwellers but long since disappeared from the American consciousness, are now resurgent, aided by increased global travel and their acquired resistance to previously effective pesticides.[18] And hospitals, where antibacterial agents are used liberally, are actually ground zero for outbreaks of Methicillin-resistant *Staphylococcus aureus*, or MRSA, a deadly and debilitating disease that has become resistant to almost all known antibiotics.[19]

Likewise, in human affairs there is abundant evidence that resources do not determine the outcome of conflicts. The 9/11 attacks, which caused untold billions of dollars of damage, not to mention thousands of lives lost, cost al-Qaeda less than $1 million.[20] And insurgencies in Afghanistan have demonstrated time and time again that even overwhelming force is no guarantee of winning an escalation.

The paradox of poorly resourced organisms adapting more quickly than their well-heeled opponents may have a simple solution. It may be simply that the best adaptations don't require many resources at all. In other words, the stake to get a seat at the escalation table is really cheap, and once you are there, you're as well-equipped as anyone to have a winning hand. This is especially true for adaptations, like the stinging cells of an anemone, that confer multiple benefits. Humans long ago developed complex cognitive abilities, so putting those into use in conflict is essentially a free form of adaptation. Moreover, we can't know the price of adaptation beforehand because we don't know exactly what the problem that needs to be solved is. It may be that an out-of-control truck is hurtling toward your car, in which case having had the resources to be ensconced in a Volvo rather than in a beat-up old Ford Pinto would likely confer a survival benefit. But it may be that your biggest worry is the potential for infection from a cut in your toe, in which case some basic awareness and wound management is all you need. Unfortunately, we often learn only retrospectively that resources were never the best answer to the security challenge that was posed. Welding plates of scrap metal to Humvees by U.S. soldiers was likely a better adaptation than waiting for the Department of Defense to roll out multi-million-dollar MRAP vehicles years later. Even better, and cheaper, was talking to local warlords to establish peaceful relationships and gain tips about bomb makers and bomb deployments.

Thinking a little more mechanistically, it may be that motivation plays a big role in overcoming a resource disadvantage. Consider a coyote chasing a field mouse. Both animals desperately want to win this interaction, but one of them is just a little more desperate. If the coyote fails, it misses its dinner, but if the mouse fails, it loses its life. This "life–dinner principle," a term coined by evolutionary biologist Richard Dawkins,[21] is believed to give an evolutionary advantage to prey species. The more honest nature films these days show the blooper reels where the lioness misses the zebra and tumbles headlong into a mud pit, revealing how often predators actually fail. This same logic applies to many human security escalations. Kenney noted that the narcos (the drug dealers) fall on the "life" side of the equation, because failing to evade law enforcement (the "narcs") will get them killed or captured, whereas the narcs may just lose a promotion or congressional favoritism if they fail.[22] Indeed, Kenney notes that narcs often use failure to justify calls to Congress for *more* funding, meaning that the relative adaptational motivation to not fail is even greater for the narcos.

But motivation can't be enough. After all, some zebras who are presumably very motivated to survive still get killed, and many more insurgents have been killed by the superior firepower of allied forces than have killed their supposedly less motivated enemies. What else keeps an organism or an organization from being overrun by its considerably more resourced competitor? Information provides a pathway for adaptation that can slip its way around even a mountain of resources held by an opponent. I have argued that information from nature is easy to obtain by those willing to act as naturalists and study it carefully, but this doesn't mean information flows freely in nature. Rather, natural systems have developed diverse pathways for using and sharing information. Secret codes, deception, and double crossing are as common in nature as in human society. The next chapter reveals some of the diverse ways that organisms in nature have come to use information and communication offensively and defensively, and what it could mean for our own adaptation and survival.

chapter seven

CALLING YOUR BLUFF AND BLUFFING YOUR CALL

THROUGH MY NATURAL SECURITY PROJECT, I met Dan B., an expert in surveillance and threat assessment. He is a critically important contributor to the project, but he can be hard to track down sometimes. Sometimes he'll be deployed in the field for months at a time in a remote outpost in the Rocky Mountains. Then I'll learn that he'd been roaming about high altitude places in dangerous territories of Pakistan, including the town where Osama bin Laden was discovered to be hiding out. From what he is able to tell me about his work, I've learned that he spends his time there stealthily deploying remotely operated drones and running a network of observers who fastidiously watch how individuals in populations respond as these threats enter their community. He wants to know exactly how a population under stress responds to a threat, and how their response changes over time. Do individuals respond in a rational or even predictable way to threats? Do individuals eventually stop caring if something once perceived as a threat never unloads with its full arsenal? He meticulously notes the responses of each individual, categorizing and quantifying their reactions with the detached air of a scientist.

That's because Dan is a scientist. Dr. Blumstein is a behavioral ecologist at UCLA, to be more precise. The remote drone he deploys isn't a spy plane from Lockheed Martin's Skunk Works but a stuffed badger pelt mounted on a customized remote-controlled truck chassis known as Robo Badger. His subjects aren't Pakistanis caught in the crossfire of the war on terror, but their marmot compatriots. And his networks of observers are students who are learning to study animal behavior.

Marmots are small alpine mammals that resemble overstuffed squirrels with stubby tails. They live in small groups that forage in alpine meadows, ever watchful for predators. Dan is particularly enthralled by marmots, even hosting a website, "The Marmot Burrow,"[1] dedicated to them, which includes, helpfully, that they make decent pets because they can be housebroken, and when you go on vacation they will just hibernate. Dan has spent countless hours quietly observing his subjects, getting to know each individual and discovering that they, like humans, have their own idiosyncrasies and their own tolerance for risk and uncertainty.

When Dan really wants to pinpoint how the threat of predation affects prey's behavior, he uses an array of ingenious homemade gadgets and sensors, including the fearsome remotely operated Robo Badger. When he drives Robo Badger into the territory of a peaceable group of small marmots, all hell breaks loose, but Dan is prepared to note how each individual responds. Using this mix of keen observations of nature and clever experiments, Dan has been able to catalog general rules of how threatened populations respond to their adversaries.

For example, in keeping with our understanding that variation is the fundamental building block of nature, Dan finds that not all marmots are equally well-adapted for survival. In particular, Dan has seen that some individual marmots, who he calls "Nervous Nellies," cry out often to the group that a predator is near and often make false alarms, while others ("Cool Hand Lukes") only make a signal when there is a clear and present danger. Unlike the parable of the boy who cried wolf, however, other marmots do not come to ignore the calls of Nervous Nellies—if they did, they would get eaten during the rare times when the Nellies (which, like the Lukes, can be either gender) are honestly signaling a predator. Rather, they are forced to spend extra time and energy (which means less time eating and reproducing) trying to figure out if the call is meaningful. By contrast, the Cool Hand Lukes, who always produce honest signals, make life easier for all the marmots. Those around Cool Hand Luke don't waste time on false alarms and are certain when he or she makes a call that they need to get to safety immediately.

The marmots tell us three important things about *information* and security. First, information use and sharing can be as essential to survival as any other adaptation. Second, both a key goal and a resultant outcome of using information in survival situations is to create or reduce *uncertainty*. Third, the way *receivers* of information—both your friends and enemies—perceive the signals you are sending is vitally important to your survival.

This chapter is about how the use and abuse of information can lead to vastly different outcomes for security. I will explore the mechanisms nature uses for information sharing and how human security systems often operate in a completely contrary way to nature.

Living organisms can keep up with evolutionary arms races even when they are vastly out-resourced, if they know how to use information correctly. Clearly, al-Qaeda has continued to operate despite overwhelming resources deployed against it by sharing information and motivating supporters on the Internet.[2] Michael

Kenney notes that drug traffickers have a natural advantage over law enforcement in that they know exactly how they will carry out their own illegal activities whereas law enforcement usually has at best only small amounts of this information.[3] On the flip side, this informational advantage can be lost quickly, and that stealthy adaptational pathway may be lost in an avalanche of resources from the other side. As Kenney notes, "Traffickers' and terrorists' primary advantage in competitive adaptation is informational: when they lose their covert edge because law enforcers have identified and located them, they face great difficulty overcoming the force advantage of the states."[4]

But using information covertly is only one way that information influences adaptation in escalatory conflicts. Public information also plays a huge role in how organizations both with and without resources adapt to security challenges. Scott Atran, who has interviewed hundreds of radical terrorists, argues that "publicity is the oxygen of terrorism." He cites Saudi Arabian general Khaled Alhumaidan, who deftly highlights the difference between the clash of physical resources and the power of informational resources when he remarks, "The front is in our neighborhoods but the battle is the silver screen. If it doesn't make it to the 6 o'clock news, then Al Qaeda is not interested."[5]

Public information drives acts that compromise security as well as our responses to those acts. In the same way that new selective forces (such as an ice age) have radically reshaped evolution and escalation throughout Earth's history, public information widely broadcast has become a new selective force in our security battles. This has both positive and negative impacts for how we adapt. The July 2010 leak of 75,000 documents pertaining to the war in Afghanistan on the website WikiLeaks revealed, in part, that many apparent successes in winning over local leaders in Afghanistan were short-lived, at best. These releases of information could be seen, alternatively, as an essential selective force for ensuring that our adaptations aren't wasting resources or missing threats to our security, or, in the perspective of a U.S. congressman who called for the execution of the suspect who leaked the documents to WikiLeaks,[6] a source of treason.

In other words, information and communication can be both an adaptational advantage and a trap. Information is cheap, but it's not free. Just the act of using information makes one vulnerable. Crafty cuttlefish have adapted around this problem by learning to communicate with one another in secret messages coded in polarized light, which their predators can't detect.[7] This allows a single cuttlefish in a group to warn the others without attracting the predator's attention to itself. But a large military operation inevitably leaks information that is much more difficult to conceal. Insurgents in Iraq and Afghanistan learned quickly to identify the signs of a changeover in troop deployments, and it is suspected that they step up their attacks just after the new troops arrive.[8] The controversial WikiLeaks episode was made possible by the increased exchange of information between intelligence agencies and the military called for after 9/11. Both al-Qaeda and regular armies, as well as narcos and narcs, reveal potential points of attack when they share information by cell phones or on the Internet.[9] A single errant phone call by Osama bin Laden's most trusted messenger was enough to help U.S. intelligence forces track down his location. Repeat offender criminals leak information each time they commit a crime related to their behavior, their methods, and the locations of their crimes. If they are not careful to continually vary how they act, they may be susceptible to new analysis methods that link crimes by the same person via the signals they send while committing each crime.[10]

Focusing on information sharing may be the most effective way to neutralize a threat. For example, scientists have recently discovered a radically different picture of the bacteria that cause diseases like cystic fibrosis. Once thought of as lone operators who cause damage when combined with many other renegades, these microbes are now known to work in coordinated fashion and use advanced communication strategies, including coordinating among themselves and intercepting signals sent by the host body's immune system. Whereas simply attacking these bacteria with antibiotics might lead to selection for antibiotic-resistant strains, targeting their *communication system* by blocking the molecules that send signals between bacteria, or those that detect signals from their hosts, is a promising new approach to treating microbial diseases.[11]

USING UNCERTAINTY TO YOUR ADVANTAGE

Not surprisingly, there are myriad ways natural systems share information, but in that reassuring way that natural systems build complexity and diversity out of simple building blocks, the wide array of natural information use coalesces around a single overarching theme: *organisms seek to reduce uncertainty for themselves and increase uncertainty for their adversaries.*

The need to create uncertainty for an adversary and reduce uncertainty for oneself and one's allies explains many complex behaviors in nature. Birds flock to cause uncertainty for a predator about any individual's vulnerability, while predators use camouflage to create uncertainty about when and from where their attack will come. This simple rule also explains the phenomena that occurred during the year I lived in Washington, D.C. That summer, sidewalks crackled underfoot like a fall day, the sky was filled with a high-pitched wail, and backyard barbecues suffered a most unusual precipitation. The singular culprit of this strangeness was the periodical cicada, a large winged insect that had emerged by the trillions after seventeen years underground. The cicadas leave their underground dens for a brief period to mate and deposit eggs in tree branches, which will hatch into larvae that drop to the ground and bury themselves for another seventeen years. Although local predators certainly feasted on this unusual treat, the cicadas' numbers literally overwhelmed the birds and squirrels, and a large number of the insects completed this vital part of their life cycle without being eaten, thus raining dead bodies onto barbecues and littering the sidewalks with their dry husks. Had the cicadas emerged in a regular period like every summer, predators could accommodate their own life cycles to match this emergence and maximize their populations to feast on the hapless insects. But these seventeen-year periodical cicadas, like their thirteen- and seven-year periodical cousins, had evolved to exploit the mathematical uncertainty of prime numbers to avoid emerging during banner years for predators.[12]

By contrast, most of the security screening we conduct tragically reverses the uncertainty rule of nature—that is, it makes life less uncertain for our adversaries and more uncertain for ourselves. When we widely advertise what we are looking for and how we are looking for it in security screening (which we must do when we order everyone passing through security to be screened in an identical fashion), our adversaries greatly reduce their *uncertainty*—they now know exactly what is being looked for and they can then work on adapting ways to get around the screening.

At the same time, we increase our own uncertainty by constantly crying "predator!" (or, as the case may be, "terrorist!"), which continually erodes our confidence that anyone really knows what is going on. Over the decade that the TSA used its color-coded Homeland Security Threat Level Advisory System—that five-color warning scheme prominently displayed in airports and other public facilities—it rarely changed the warning level. In airports, the threat level stayed constantly at orange from August 2006[13] until the program's demise five years later. This was not a convincing show to our enemies that we actually did know what they were up to, and it also didn't give clear information to the population it was supposed to protect. On a typical "orange day," video screens under the threat-level announcement told us that we should "Report Suspicious Activities to Authorities." This didn't seem like information particularly targeted to a highly elevated threat level—after all, shouldn't we *always* report suspicious activities to authorities?

Like the Nervous Nellie marmots, these kind of ambiguous warning systems don't work because they create uncertainty among the population they are supposed to protect. And uncertainty, in turn, creates stress. Robert Sapolsky, a neurobiologist at Stanford University who is recognized as one of the world's experts on stress, began to recognize this during his first studies of baboons. He found that individuals who were low on the social ladder experienced much more uncertainty about when or by whom they would get attacked, and had much higher levels of stress and associated maladies.[14] Stress affects people in uncertain environments in similar ways. In fact, a recent medical report demonstrated that more people have died of heart failure due to increased terrorism-related stress since 9/11 than died on 9/11 itself.[15] Not all of this additional mortality is due to ambiguous signaling, of course, but uncertainty is a major component of the additional stress these people reported.

HONEST AND DISHONEST SIGNALS

One of the key ways of managing uncertainty is through signaling. All sorts of animals, as well as many plants, use some form of signaling to help themselves and their fellow individuals survive in a risk-filled environment. Signals are certainly shared between animals of the same species. They might be warnings about a nearby predator or boasts about one's sexual fitness or subtle clues about one's perceived spot in a dominance hierarchy. But there is also a lot of signaling between species. Often, interspecies signaling takes the form of prey species giving warning signals to their predators. Why would a prey species make itself known to a predator by signaling its whereabouts? Mostly to ruin the element of surprise. Predators and prey are locked into an intense and highly evolved game. But as I mentioned in the previous chapter, most predation events result in failure. Chasing down a fleet animal, swooping down on a tiny animal from high above, and striking fast enough to immobilize a prey of equal size are hard to do under the best circumstances. When a predator that relies on surprise is discovered by the prey, it will usually call off the hunt before it's begun.

Signals do not even have to be perfect to work. Many benign animals mimic the coloration of a similar toxic animal to trick predators into avoiding them. Biology textbooks tend to show side-by-side pictures of the most perfect mimics, but in reality there are many mimics that aren't nearly so perfect. For these animals, it is just enough to confuse a predator long enough to get a head-start escaping.[16] Sometimes, in fact, mimicry that is too perfect can cause trouble. It was recently found that two tropical butterfly species, which are near-perfect visual mimics of one another (an adaptation to deter predators, as their coloration signals that they are poisonous), end up confusing males, who spend considerable time determining if the female they are courting is really of their species.[17] Cuttlefish, intelligent cousins to the octopus, are masters of camouflage, but not necessarily because they are perfect at it. For example, when they are hiding in sea-grass beds, they don't try to match the patterns of grass using their color-changing skin; rather, they raise their arms, swaying them in synchrony with the ocean currents. If they had tried to perfectly match the sea grass, even small differences would leave the animal exposed—but by adding more stripes (their arms) to the striped environment of the sea-grass bed, they create a disruptive pattern in which it is difficult for a predator to be sure what is an edible creature and what is inedible background.[18] This imperfect but advanced form of camouflage has drawn the attention of military analysts who recognize that dynamic, changing patterns are far more effective than the static camouflage currently employed by the military.

Humans have occasionally employed imperfect signaling in successful improvements to their security. When desperately short of resources in the face of the German *blitzkrieg,* Britain created inflatable dummy tanks and airplanes that both diverted German bombs to harmless targets and, while not perfect, distracted the enemy enough to boost wartime production of real tanks and airplanes.

Imperfect signals serve a purpose in some cases, but Britain wouldn't have held off the blitz with inflatable weapons alone. Dishonest signals are quickly exposed when they come into contact with the signal receiver. Consider the Argentine jail that placed a mannequin of a security guard made out of a soccer ball and a guard's cap in one of its watchtowers to save money under a tight budget. Two prisoners promptly escaped once they figured out the ruse.[19]

Organisms in communities need the Cool Hand Lukes—the truly useful signals—accurate, directed, and clear. Dan Blumstein has compiled a number of examples of "honest" signals from nature. Male red deer, for example, roar to impress females and express their dominance to other males. It has to be an honest signal because the depth of the roar is controlled by the size of the deer's body, which in turn is a fairly reliable indicator to the deer's potential mate and potential competitors of his fitness, or his capacity to grow large. Male peacocks also make demonstrably honest signals to females. The more energy they are able to put into their display, the more opulent and bright their feathers become. A malnourished peacock, which would make a less fit mate, cannot put much energy into making a flashy signal.

Marmots also tell us that to be effective, signals better be perceived properly by their intended targets. If Dan used Robo Badger in a truly deadly fashion against the marmots, even Cool Hand Luke's best alarm calls directed at the erstwhile badger would be useless because the remote-control car powering the badger pelt does not hear. From lowly invertebrates to otherwise unremarkable mammals like ground squirrels, we are beginning to discover that animals do indeed target their signals very accurately toward their receivers. Cuttlefish make very different signals when faced with different predators. If the nearby predator is a fish, which tend to assess a potential prey item quickly and then move on, the cuttlefish puffs itself up as big as possible and uses its color cells to form two "eye spots" on its back—an effective bluff that is also utilized in a more permanent fashion by less-flexible creatures such as the four-eye butterfly fish. But when the cuttlefish is facing off against crabs, which hunt by smell, or dogfish, which use electric fields to hunt, such visual displays won't work, so the cuttlefish just jets away as quickly as possible.[20] Ground squirrels were recently discovered to also possess a fine-scale sense of how predators received their signals. Squirrels will make shrill alarm calls aimed at bird and mammal predators (who can hear), but when faced with a snake (which doesn't hear), they don't call but rather puff up and shake their tails. If that snake happens to be a rattlesnake rather than a gopher snake, the squirrel also *heats* up its tail because rattlesnakes are unique in that they perceive infrared radiation.[21]

We have not always abided this lesson in our security efforts. For example, shortly after Osama bin Laden was killed by U.S. Special Forces, the United States released captured videos of bin Laden's life in hiding. The Western press took the signal exactly as the government intended, discussing the videos of the bedraggled bin Laden bobbing his head like a mentally deranged man on the floor of a sparsely furnished room under headlines like "Videos Demystify the Osama bin Laden Legend."[22] Yet the video signals were received in a completely different way by the radical Muslim world. They saw images of a very rich man nonetheless living with few possessions. They saw that he kept his long beard despite the need to disguise himself, and that he wrapped himself in a dark blanket and prayed on the floor, bobbing his head in a ritualistic way to what was likely a reading of the Koran (sound was edited out of the released video), all signals of a devout Muslim following in the way of the Prophet.[23] What was meant to show an image of a weakened al-Qaeda leadership actually signaled the complete opposite to potential followers.

Those plucky Brits in World War II do provide effective examples of signaling directed precisely to the receiver to improve security. R.V. Jones recalls in his autobiographical account of the efforts of Britain's Scientific Intelligence Service[24] that in order to disable German radar, Jones and his ragtag team of scientists needed to figure out how it received signals. Once they were able to do this, they developed simple devices like strips of reflective fabric cut to just the right length that could be dropped from aircraft so that German radar would see not a squadron of bombers, but a harmless cloud.

Signaling is essential to security, but it is not without vulnerabilities. Signals can overestimate or underestimate the true risk of a threat. They can be perfectly correct in the information they give but be so far removed from the source of the threat that they don't seem particularly relevant to the receiver. Or they can be so repetitious that either the receiver becomes habituated to them and they foment a much smaller reaction or they annoy the receiver to the extent that the signal is shut down if possible.

There seems to be a consistent reaction to repetitious, unchanging signals: checking out. Dan Blumstein, the marmot lover, and ecologist Elizabeth M. P. Madin investigated this phenomenon, what behavioral ecologists call "habituation," by examining visits to the Department of Homeland Security's Ready.gov website and calls to the DHS information line—which is supposed to inform civilians about what they should do to be prepared for homeland security emergencies—in relation to the threat level. They also looked at public opinion polling data to see how worried people reported feeling about terrorism. They found that seeking information about terrorism decreased over time and had no relationship with the threat level.[25]

Egypt cleverly exploited the habituation effect of signals in its conflict with Israel. In the weeks preceding the 1973 Yom Kippur War, Egyptian forces conducted forty military exercises in plain view of Israeli forces along the border. Israeli sentries had come to expect the movement of armed soldiers and armor in the region. The first morning of the war started as many of the preceding days, with Egyptian military exercises, but this time the Egyptian forces kept right on moving across the border and initiated a "surprise" attack on Israel that was planned in plain view for weeks.[26]

Egypt turned a pernicious problem of repetitive signaling on its head to create a security problem for its enemy. Most of our signaling technology, while designed to keep us safer, in essence provides the same repetitive habituation, and ultimately makes us more vulnerable.

An ever-present, never-changing signal is about as effective at keeping a vulnerable population alert as a diorama of a grizzly bear is at scaring a museum visitor. But we certainly all jump when an infrequent signal, such as a fire alarm, suddenly goes off. Nonetheless, as I've already mentioned with regard to the tsunami alarms stoned to death in Banda Aceh and the millions of disabled smoke detectors in American homes, most alarm systems don't provide us with the security we would like them to give us. Likewise, explosive gas alarms on the *Deepwater Horizon* rig were disabled months before the deadly explosion there. Collision alarms on the Washington, D.C., metro system were routinely ignored before a deadly crash on a red line train there; air traffic control alarms have been ignored, leading to crashing planes[27]; and Vietnam-era fighter pilots, who suffered high casualty and capture rates, shut down surface-to-air missile alarms because they just had too much information coming at them in the cockpit. It's not like any of these alarms are warning about trivial matters. Nor is it the case that these alarms *don't* save lives. One study from Ontario, Canada, showed that nearly 30 percent fewer deaths occurred in house fires where a fire alarm was present and working compared to those homes without working fire alarms, whereas U.S. reports place the percentage of fewer deaths in homes protected by working alarms at about 50 percent.[28] Most of these alarms are disabled or ignored because they produce too many false alarms.

So how can alarm systems work better? Re-sampling—in other words, bringing in a redundant element—can help a detector determine if the original stimulus was in fact a real threat. Mike Dziekan reports on a company that had an optical fire detector installed above the camera that made their photo IDs. One time when a new ID badge was being made, the flash from the camera triggered the alarm, causing the whole facility to be evacuated. If the alarm simply sampled the environment multiple times it would have caught the false alarm.[29]

Marmots and other animals in groups are able to use a form of re-sampling to ensure that they are getting honest signals. They can listen to the alarm calls of several individual marmots to determine if a threat is real. If only Nervous Nelly is making alarm calls, there is a much smaller risk than if Nervous Nelly and Cool Hand Luke and other marmots are making a call. Likewise, as I mentioned in Chapter 5, honeybees check multiple sources of information, essentially counting votes of many individuals in the colony, before making a decision such as the need to move the hive because of imminent danger.[30] These represent the same type of adaptive redundant features discussed in Chapter 5. Obviously, a key factor in all of these redundant information systems is speed—they won't work if taking multiple samples for the presence of predators or the presence of smoke takes a long time.

But even the most robust signaling strategies can be overcome. When an enemy learns your signaling pattern, it can use it against you. A U.S. soldier can salute, wear a uniform, and march in line to signal his loyalty, but it doesn't mean he is loyal. This contradiction was tragically seen when U.S. Army Major Nidal Malik Hasan killed thirteen people and wounded thirty others on Fort Hood Army base in Texas on November 5, 2009. Hasan was shot four times by a soldier on the base, ending his shooting rampage. Although he barely survived, it would seem likely that an attack of this type at a military base was a suicide mission, indicating both the dishonesty of his signals of loyalty to the United States and the honesty of his signals of loyalty to radical Islamic ideals.

Suicide attacks are the ultimate expression of an honest signal—in this case, of commitment to a cause. Several of Hasan's truly honest signals were known through his idiosyncrasies—he once launched into a spirited defense of radical Islam and suicide bombing during what was supposed to have been a medical lecture, and he communicated with Anwar al-Awlaki, a radical Muslim lecturer[31]—but these signals were not interpreted by his superiors as overriding his more routine signals of loyalty to the U.S. Army.

YOUR FACE GIVES YOU AWAY

Criminals like Hasan may be able to use a number of superficial signals—their rank and uniform—to send dishonest signals, but their biology and their biological legacy send far more honest signals out, and we need to become better at detecting those signals. In particular, humans give away a lot of information in their facial expressions and body language. Even just the shape of your face may reveal your level of aggressiveness. A study of hockey players showed that the larger the width-to-height ratio of a player's face, the more likely he was to have accrued penalty minutes for fighting and other aggressive transgressions.[32] It's believed that this wider facial structure is linked to higher testosterone levels during development and thus higher aggressiveness.

Humans also have a very strong capacity to discern this signal: people representing several different cultures were able to accurately assess upper body strength (also a proxy for testosterone levels) of strangers even when they could only see their faces.[33] But even the toughest hockey players are typically not engaged in terrorist activities (although a heavy hitter like Alexander Ovechkin certainly terrorizes his opponents on the ice), so while unchangeable features such as facial structure may be a subtle clue to the personality of our fellow humans, they don't give us the fine-scale signal of nefarious intent that we would need to identify who is *immediately* dangerous.

Using behavioral traits to screen people entering secure areas exhibits multiple features of natural security systems. Like other social organisms such as wasps, wolves, and primates, we have a highly evolved ability (including brain specialization) for facial and behavioral pattern recognition.[34] Crows, who are also highly social, can remember for their entire lives the face of a human that has mistreated them when they were younger, and they may even share this learning among their social group.[35] Behavioral recognition is a redundant defense effective against multiple types of criminals (e.g., drug smugglers or terrorists) even if their planned crime involves no concealed contraband. It can operate independently of racial profiling—Darwin, through observing many different cultures during his voyage on the *Beagle,* recognized that there were common expressions across human cultures and between different animal species as well.[36] Finally, behavioral screening returns control of uncertainty to the population it is trying to protect because it can be conducted from hidden vantage points or video. As a head behavioral screener at Dulles Airport (one of 161 airports where behavioral screening has been deployed by the U.S. Transportation Security Administration[37]) remarked, "The observation of human behavior is probably the hardest thing to defeat. You just don't know what I am going to see."[38] The efficacy of layering discrete behavioral screening with other levels of verbal and nonverbal intent detection systems is currently being investigated.[39]

Behavioral screening is also not a perfect system. Scientists and the Government Accountability Office bemoan the fact that the TSA's behavioral screening was deployed with very little scientific evaluation.[40] They would like to see more experimental testing, but I'm skeptical. As an ecologist studying large-scale ecological changes like climate warming and fisheries collapses, I am increasingly convinced that many problems can't be ethically or effectively studied with a manipulative experiment. Just as we can't design an experiment to replicate the incredibly complex suite of changes that climate warming will bring—we have to rely on detailed, real-world observations of how warming has already affected the planet and then project forward—it's hard to imagine an experiment that can study the effectiveness of behavioral screening before it's implemented in a real-world setting. Probably no actor is good enough to mimic the human physiological changes and outward expressions of those changes that occur when the mind and body are under extreme stress, as a real-world terrorist or drug smuggler would likely experience. To really study behavioral screening, you have to set it up and carefully monitor how well identification of certain traits leads to arrests of criminals.

Other critics argue that behavioral screening just hasn't worked. *USA Today* recently reported that sixteen passengers evaded behavioral screening and later turned out to be linked to terrorist plots.[41] But this shouldn't be surprising. Humans are very malleable. If these people, who were later linked to terrorist plots, were not actually involved in a plot at the time they went through behavioral screening, it's likely that they didn't express the outward signs of concealed behavior. Behavioral screening is only one part of a multi-layered screening system.[42] If these sixteen passengers were truly a threat to air travel, they should have been identified by intelligence operations and placed on "no fly" lists before they ever got to the behavioral screening phase.

Still other critics argue that behavioral screening works too zealously, pointing out that many travelers are nervous about flying and likely to exhibit tense behaviors. This argument is backed up by the fact that of the 152,000 or so people singled out for secondary screening at airports where behavioral screening was used between 2006 and 2009, only 1,100, or less than 1 percent, were subsequently arrested, with critics pointing out that this is a low positive rate for the number of passengers singled out. There is an undertone of indignation (and some pending lawsuits) on behalf of the violated civil liberties of these unfairly targeted people. But the reality is, all of us are unfairly targeted at airports; from the moment we walk in under video surveillance to the time we shed our shoes and subject our luggage and bodies to X-rays, we are presumed guilty until shown to be innocent. For the most part, you check your civil liberties at the curb with your luggage.

A wider look at the numbers weakens the arguments of the critics. Consider that roughly 2 *billion* people traveled through these same airports during the study period, meaning that less than 0.01 percent of passengers were detained somewhat longer than the normal delays for screening that everyone receives. If we expand the view even more to include arrests of all U.S. air travelers (including airports that don't use behavioral screening) at screening checkpoints using the best numbers I could get from TSA, we find that 855 people were arrested at security checkpoints in 2008–2009,[43] a two-year period in which roughly 1.4 billion people travelled in U.S. airports.[44] This general arrest rate—0.0000006 percent—makes the 1 percent arrest rate of people detained through the more targeted behavioral screening seem like a pretty effective process after all.

Going from animal signaling to the unique facial and behavioral attributes of humans begins to close the loop. We started by looking at nonhuman natural systems that give us some insight into how humans might design and implement security systems. Now we are looking directly at what human nature can tell us about security failures and opportunities in human societies. In the next chapter I further blur the lines between nature and human society through an exploration of how human nature and the belief systems that we develop as humans growing up in different environments become the source of many of our security problems.

chapter eight

THE SACRED VALUES OF SALMON AND SUICIDE BOMBERS

Salmon go to great lengths to kill themselves. After a short few years frolicking in the open ocean, they may travel thousands of kilometers to get back to the precise stretch of the same river in which they were born. On this journey they will have to slip past the birds, bears, sea lions, and humans that gather at river mouths to feast on them. They must swim exhaustively upstream for many miles, using most of their energy reserves to leap up waterfalls or swim ladders (artificial waterfalls constructed on the sides of artificial dams) until they reach their spawning grounds, where their last gasps are spent producing eggs or fertilizing them with sperm before collapsing in death, never to see the ocean again.

From an evolutionary standpoint, it's not hard to find sense in these suicide missions—the salmon are passing on and multiplying their genes in a habitat that has already been proven (by the adult salmon's own experience) to produce strong and reproductively fit salmon. People tend to admire the determination of the salmon. At the very least, we generally don't call the salmon "irrational" or "crazy" for their journey. We do, however, freely launch those pseudo-psychological assessments on human suicide bombers. Yet salmon and suicide bombers are not as different as their outward appearance would indicate. The most important difference between them is neither fins versus arms, nor gills versus lungs, but that the salmon (despite the dams choking up the rivers) still lives in the environment its ancestors evolved in for thousands of generations, while the suicide bomber does not. I'll argue in this chapter that suicide bombing is just an extreme case at the far end of a spectrum of behaviors related to establishing and reinforcing self-identity that impart survival to organisms.

Those behaviors tend to be the things that many people would call irrational beliefs, and they turn up all the time in security questions. Pakistanis and Indians fight a seemingly endless war with potential for mutual nuclear annihilation over a narrow strip of barely habitable territory. People stay in their homes despite clear warnings and even as the floodwaters rise into their attics. And we ignore over 100 years of collected scientific wisdom while we watch human-induced climate change alter our entire planet.

The naturalist's view on security doesn't allow us to simply label something "irrational" and then dismiss it. Just as a biologist wants to get to the root of what makes a peacock grow such outlandish feathers or an immune system suddenly turn on its own host's body, a natural-security approach tries to get inside these behaviors that compromise our security, tracing their roots back as deep in evolutionary time as possible and figuring out what they mean in today's society.

Evolutionary psychologists, who study the ancient roots of modern human behavior, argue that religious fervor didn't develop in the modern world but in a world completely unlike the one we have briefly inhabited now.[1] In this early world, humans lived in small close groups that struggled constantly to obtain enough resources to survive. Only rarely did they encounter small groups of other humans, and if their interaction wasn't about trading resources, it was likely because one group was trying to take the other group's resources by force.

Yet almost all political analysis of human behavior tries to explain it within the narrow confines of the immediate sociopolitical environment. Some public commentators try to get us to broaden our thinking. Journalists try to remind the most short-sighted among us that there were clear signs of terrorist activity against Western targets years before 9/11. Historians admonish us to open our eyes and look at the thousands of years of history in places like Iraq and Afghanistan.[2] Political scientists urge us to look at individual security crises within their global context. I fully support these viewpoints, but I suggest that analyses digging back ten, a hundred, or even a thousand years must be nested within a perspective that goes back orders of magnitude deeper into human, and biological, evolution.

If we convert our years as humans on Earth to words in a book, analyzing security only in the context of the past few thousand years of human history is like trying to understand all of *War and Peace* by reading only the last word. If we don't understand the true nature of human behaviors we will only, at best, address the most superficial manifestations of deep-seated hatred, mistrust, and suspicion. At worst, by trying to eliminate behaviors (such as religious belief systems) that we fear but don't properly understand, we will create a rich environment for strengthening and replicating them.

THE VIRAL ROOTS OF IRRATIONAL ACTS

In fall 2007, angry mobs marched in the streets of Khartoum, Sudan, barely held in check by soldiers wielding machine guns. The mobs weren't demanding higher wages or an electoral recount, but swift and brutal justice for English school teacher Gillian Gibbons, who was standing trial that day. Her crime? She had allowed her class of seven-year-olds to name the class teddy bear Mohammed. To Western observers, the reaction by the people on the streets and the religious court, which ultimately spared Ms. Gibbons forty lashes but gave her fifteen days in prison, was nothing short of crazy.

Yet there are plenty of seemingly irrational beliefs held by groups in the Western world as well. Most biology professors at one point in their teaching careers find themselves in the tricky spot of debating a student whose deeply held religious beliefs puts him or her at odds with observable features of the natural world. This is particularly evident in the case of teaching evolution, but it seems to pop up all over the place. My moment came early in my teaching career at California State University at Monterey Bay, a public school that benefits from an incredibly diverse student body—in socioeconomics, cultural background, and even age of the students. With this diversity comes a wide range of educational backgrounds, including students with virtually no practical scientific experience prior to attending college. In a very animated lecture on the scientific method in which I expounded with great vigor the virtues of using careful observations of nature as the verifiable basis of scientific study, a student stopped me dead in my tracks by asking, "Well, what about miracles you can see right before your eyes, like the Eucharist?"

"The what?" I said.

"You know, the *Eucharist*. When the priest turns bread and wine into the body and blood of Christ?" he replied, surprised at my ignorance.

The idea that a twenty-something college student would still believe every Sunday that he was actually witnessing the blood of Christ being squeezed from bread was so far beyond my naive notions of what college science students thought about, I was momentarily speechless. I muttered a somewhat lame retort to the effect of a scientist needing to question even the most careful observations, but I was treading lightly, trying not to anger or embarrass this student in front of his classmates. Moreover, at that point, I really didn't have at my disposal a way to reconcile two very different ways of observing the world.

Scientists and social commentators have struggled with this issue in very public arenas as they debate whether evolution and "intelligent design" should be given equal weight in science classes, or how much a role religion should play in politics. Some have taken a very hard line, systematically deconstructing what they call The God Delusion (evolutionary biologist Richard Dawkins) and assuring us that "God is not Great" (political commentator Christopher Hitchens).[3] Others, like the biologist E. O. Wilson, who grew up in a Southern Baptist tradition, urge a more reconciled view of mutual respect between scientific and religious viewpoints.[4] But debating God away, or recognizing, as I try to do in my biology classes now, that some things are a matter of evidence and hypothesis testing and others are matters of faith, does not get at the clearly observable phenomenon that many people, probably the vast majority of people, hold beliefs that are patently false. More important, human behaviors that seem at first sight to be merely irrational turn out to be at the core of many security problems we face.

Oddly, a virologist, Dr. Luis Villarreal of the University of California–Irvine, has made some key discoveries about human belief systems. It is through his high-profile Center for Virus Research that Villarreal began to trace how far back in evolutionary time belief systems of some sort can be traced, but it was through his community service that he learned first-hand how strongly humans hold onto even irrational beliefs.

Villarreal is a product of the lower-income Latino community of East Los Angeles, growing up with no relatives or mentors who attended college. As the only member of his high school class who went to graduate school, and then on to a successful university career, he took it upon himself to mentor other Latinos who overcame similar social and economic hurdles as he did to make it to UC Irvine, one of the nation's premier medical research universities. As director of the Minority Science Program at Irvine, Villarreal developed a self-styled "science boot camp," in which the beliefs held both consciously and unconsciously by his students are revealed and tested against the rational construct of science. But through the years he became increasingly frustrated in his attempts to disabuse his charges of their scientifically indefensible beliefs—that the world was created in six days, that lighting a candle for *La Virgen* can bring prosperity, that every Sunday in church a priest could squeeze the blood of Christ from a loaf of bread. Although Villarreal can claim some success in terms of getting students to understand a scientific process, he found to his surprise that no matter how tough he was on his students, or how clearly he outlined the irrationality of their beliefs, many still held them dearly even as they continued their studies in sciences. The difficulty he observed in disabusing highly intelligent and motivated students (indeed, 96 percent of UC Irvine students graduated in the top 10 percent of their high school class[5]) of even the most patently falsifiable beliefs made him want to dig deeper, to uncover the root causes of these beliefs. His search was motivated by a simple question: "Why do people hold beliefs, even obviously irrational beliefs, so strongly?"

Having been an outsider to the mostly white world of academic medicine himself, Villarreal suspected on a personal level that his students' beliefs were a way to define their identity within a group and against another group (the nonbelievers) and that students put into a situation where they were the overwhelming minority would be especially inclined to tightly grip their belief systems.

But it was Villarreal's professional viewpoint, working with viruses that are among the oldest organisms on Earth, that gave him the necessary perspective to unearth the deep and tangled roots of human behavior. The evolution and development of viruses, it turns out, is inextricably tied to almost every major evolutionary advance, including the rise of modern humans, in Earth's history.[6] This means that as opposed to, say, anthropologists (who have their own important role in studies of conflict), or animal ecologists (who also contribute a key perspective on how to use information in security situations), Villarreal could deftly skip around the kingdoms of life as well as time-travel through vast spans of Earth history in his search for the origins of belief systems.

What Villarreal emerged with is a synthesis that traces the origins of *human* belief systems back to the earliest life forms, such as bacteria.[7] The exact forms of these belief systems obviously differ between, say, a bacterium, a salmon, a chimpanzee, and a suicide bomber, but the mechanism is the same. In Villarreal's theory, belief, as we know it in humans, is a form of addiction. And addiction in its pure form, according to Villarreal, is one of the oldest processes of self-preservation on the Earth, traceable to the earliest invasions of bacteria's genetic material by viruses.

Although viruses seem to cause chaos in our daily lives—at the least, they cause sick days and frantic parents rearranging day care schedules, at worst they lead to epidemics that kill millions—the virus itself wants stability more than anything. In this sense, a virus is like a businessman trying to maintain a steady clientele. More particularly, the virus is like a drug dealer trying to develop a clientele of hard-core addicts. It does this by offering protection to its clients, something like a safe place to shoot up, shielded from the police or other junkies. This safe place is created by paired genes—called an addiction module by Villarreal—that the virus inserts in the bacterial genome. One part of this pair (called the toxic gene) is destructive, killing all entering foreign bodies at will. If this gene was left to its own devices, it would destroy everything, including the host bacterium's genome itself. So it is paired with a counterpart (the *anti*toxic gene) that confers immunity to the host. This simple opposing pair—aggressor and protector—provides a way to distinguish, even in the most basic organisms, *self* from *nonself.* If something is "self," coming from the host's own body or genome, the antitoxic gene allows it to reproduce. If something is nonself, a foreign invader, the toxic side destroys it. It's easy to see why the bacterium, or indeed any other organism, would get addicted to this product pushed on it by the viral parasite—without it, any number of invaders, including the virus itself, could destroy the bacterium.

The story would end there with bacterial addicts if it wasn't such a good system these viral pushers set up. When biological systems emerge with an idea that works, it gets made again and again. Sometimes the idea is replicated exactly; thus we humans have major components of our genome (especially those vital to survival) that are nearly identical to goats and fiddler crabs and even those earliest viral parasites. But often times, good ideas are merely mimicked, taking on different forms for different organisms in different environments, even as they maintain the same basic function.

A way to detect self from nonself is one such really good idea in biology. Nearly all organisms benefit from such a system. It allows them to identify who is likely to share their interest in producing common offspring and who is likely to disrupt that chain of genetic descent. It allows them to distinguish who to school with and who to swim from, who to eat and who to eat with. Even below the level of organisms, self–nonself identification is essential. In species where females mate with multiple males, the seminal fluid around the males' sperm has evolved to protect its own sperm and destroy the sperm of a rival male.[8]

As organisms get more complex in their behaviors, they need ways to identify potential mates and potential enemies. They need ways to assess a competitor's intentions. They need ways to make friends and influence others. Villarreal argues that the same basic addiction system—a system that confers simultaneously both protective and destructive powers—fulfills all these complex needs of biological organisms.

Take the suicidal salmon. Young salmon cue into the precise chemical cues in their home stream. Then they make their way out to sea, traveling thousands of miles over two years or more, before returning to the precise part of the same stream in which they were born, in order to mate. While they may navigate by ocean currents and stars and magnetism in their open water phase, what gets them back to that precise stream riffle where they were born is the smell. Salmon possess a remarkably effective chemical-sensing organ called a vomeronasal organ (VNO). Villarreal argues that the VNO system is the same type of addiction module as the toxic/antitoxic gene pairs in viral-bacterial interactions. Indeed, a VNO system is another one of those evolutionary success stories that gets replicated in animals as different as salmon, snakes, and shrews.[9]

For salmon, a sense of self and a sense of place are inexorably linked. Any particular salmon is literally defined by its home stretch of stream. In the salmon's VNO system, home-like smells are intensified in the system and honed in upon, and non-home-like smells are rejected and effectively ignored. As a result, the salmon will relentlessly target their home spot, past anglers' hooks and gaping sea lion jaws and enormous concrete dams with their artificial fish ladders as a small (and only partially effective) concession to the salmons' unyielding will. What we admire as the incredible determination of the salmon is exactly the nature of self-identity addiction. The high threshold of acceptance into the "self" category ensures that only the most fit will survive and reproduce. This addictive system, by the time it appeared in its particular form in salmon, already survived billions of years of relentless natural selection. What are some scattered predators or concrete barriers in relation to that track record?

Like the salmon VNO, our own behaviors have driven us to do remarkable things. Our behaviors allowed us to cooperate in complex ways and form strong groups, bonded for life. In small, clever groups whose members had a deep intimacy and mutual understanding and specialized in different tasks, we pulled through any number of forces—predation, bad weather, changing climates—that could have easily wiped out our weak and nearly naked bodies.

For salmon, group survival comes in part from a common set of olfactory cues that urge the fish to simultaneously migrate to natal rivers and spawn. But humans don't have such a great sense of smell. The popularly bandied idea that invisible pheromones control our behavior, not to mention the endless iterations of supposed pheromone products purported to "drive women wild with desire," appears to have little backing in olfactory science.[10] While smell plays a subtle and not completely understood role in human mating,[11] smells don't play the dominant outward role in human identity. That is because higher primates and humans essentially turned our VNO systems off. The genes that form the VNO system are still there, but they don't get activated. Those genetic changes have obvious outward manifestations. Have you noticed, perhaps while walking your dog, that we humans don't scent mark or eagerly sniff one another's nether regions when we run into a friend on the sidewalk?

But we do mark territory; just look at the graffiti scrawled across the walls in the tough neighborhood where Luis Villarreal grew up. That written marks were substituted for scent marks is a clue to the force behind our current sense of identity. Written symbolic language, which recent reexamination of the earliest cave paintings suggests may date back, not three or four thousand years, but perhaps as long as 30,000 years,[12] is a uniquely human attribute and one that codifies our identities—especially our group identities.

Written language has a key role in codifying religious beliefs. As Villarreal points out, the word *literate* originally meant "one who can read holy scripts." Not only are religious beliefs often spelled out in written tomes, but religious myths also contain curious references to written materials. God doesn't just tell Moses the Ten Commandments; he gives them to Moses in written form on stone tablets. And when Moses grows angry with the Israelites for their idolatry, he smashes the tablets as a symbol of the broken bond between the Israelites and their one true God. The deference to written scripture goes beyond Judeo-Christian religions as well. A well-respected Japanese Shinto group, Oomoto, was codified in the late nineteenth century when Deguchi Nao, a supposedly illiterate housewife, suddenly had a vision that she transmitted into calligraphy that she scrawled across the walls of her cottage.[13] This is not to say that nonliterate cultures can't develop religious beliefs, but rather that written language provides a powerful symbolic shorthand for ideas that defy observable natural phenomena.

Defiance in the face of observable evidence is something that continually baffles outsiders trying to understand behaviors of individuals in tightly bound human groups—be they scientists trying to debate creationists or CIA agents trying to understand why someone would blow himself up for a cause. The rationalist-evolutionist deftly dismantles the structure of creationist theory with a few pieces of devastatingly incontrovertible evidence, but then can't understand why the school board (freshly stocked with evangelical Christians) votes to "teach the controversy" in her daughter's public school. People coming from this rationalist perspective tend to think that the resistance to rational testing of ideas is a *weakness* of religion—when in fact the opposite is true. Religious beliefs, perhaps more than other human belief systems, function well as a strongly addictive system because they substitute symbolic group identification for any type of rational-based test of group fidelity. The core ideas of religious conviction are universally true to believers and will remain so as long as adherence to religious laws is maintained, regardless of what some egghead scientist or analyst says.

Indeed, the high bar of irrational thought associated with most religions is a selective force that increases the strength of the belief system through time. Stream reaches that require salmon to make large leaps of gravity to get home and religions that require large leaps of faith for acceptance into the sect both enrich their populations with individuals that are especially capable of making these leaps. In part, this is an example of "honest" or "costly" signaling as discussed in Chapter 7—there is no bluffing your commitment to the group if you will injure or kill yourself on its behalf.

Joseph Henrich argues that in humans, the value of such costly signaling is reinforced by language and a complex mind.[14] That is, I can use words to trick you into thinking I'm a member of your group, and I can do it without much cost to myself (talk is cheap). But you know I can do this, so if you really need to trust me, you're going to need something more than words. Actions are much harder to fake, and they require a real cost, which could be time (the pre-teen studying for her bat mitzvah), physical pain (American Indian sun dances where the young initiate is held to a post in the hot sun and subject to ritual piercing), or even ostracism from another part of society (polygamist sects that require women to dress in antiquated costumes and keep their hair in decidedly unfashionable styles).

An important corollary to the costly signaling hypothesis is that it does more than just confirm the identity of an individual to the group. Rather, individuals that engage in costly behaviors for a group serve as recruitment tools to attract new members. And among existing members, costly behaviors encourage those with only weak commitments to strengthen theirs and become more deeply involved. Both ancient Christians and modern radical Muslims have been recruited to these groups after witnessing acts of martyrdom by avowed group members.[15]

HOW DO YOU SOLVE A PUZZLE LIKE RELIGION?

Just because they are deep-rooted does not mean belief systems are necessarily locked in forever. Certainly, we're able to trade more primal evolutionary signals for modern ones. That is why a short, nearsighted, balding weakling, who would have been an evolutionary dead end in our hunter-gathering days, may still find a fine mate, especially so if he drives a Ferrari. If a modern human female can calculate that the resource-gathering ability of the Ferrari driver may make up for his obvious physical weaknesses, so too can a modern Israeli or Palestinian realize that coming to the negotiating table with an eye to the future rather than to the insults of the past will lead to a much better future than engaging in escalating acts of violence. A modern jihadist can recognize that continuing his education, learning new skills, and getting a mainstream job will give him a far better chance of propagating his genetic code than committing an act of martyrdom. Still, many do not, and it would help to understand why they do not.

In order to do so, we need to recognize that even deep-seated adaptations are constantly under evolutionary pressure. Evolution doesn't just produce one-time solutions that either survive or fail. Those that survive do so both because they were good solutions and because, through modification, they adapt to new situations. So, it is not enough to track down the evolutionary roots of a behavior; we also need to see how that behavior has adapted in the modern world.[16]

Detailed ethnographies of some people who subscribe to some of the most violent belief systems are illuminating how belief systems manifest in the modern world, and are supplying viable pathways toward diffusing the most dangerous aspects of these beliefs. When biologists study nature we typically ask questions and channel our voice through nature to answer those questions for ourselves. Fortunately, with humans, no such ventriloquism is needed. We can ask humans about their beliefs and get an answer directly from them. True, humans are clever enough to obfuscate in their answers, replying in half-truths and ambiguous language. But in reality, we can account for these human biases in the same way we account for biases in nature—we don't only "ask" one starfish what color it is and then declare that all starfish are that color. Likewise, by being conscious of human biases and designing questions and experiments carefully and, most of all, applying these treatments to many, many people of different backgrounds, one can start to get a sense of general patterns in human beliefs.

Scott Atran, an anthropologist who studies people who hold certain beliefs with such tenacity that outside observers would likely call them irrational—young men who vow to commit acts of suicide terrorism, radical right-wing Israeli settlers, and militant

Hamas members—calls these apparently intractable beliefs "sacred values." Examples of sacred values include Palestinians' belief in a right of return and Israeli settlers' belief that they were chosen by God to live on the land. Through thousands of interviews with Israelis and Palestinians and their leaders, Atran has begun to identify and characterize sacred values, and even develop strategies for resolving conflicts built around them.[17]

One of Atran's earliest conclusions was that sacred values are not fungible. Sacred values have been shown to hold up to all sorts of proffered alternative ideas. They resist compromise, they resist financial incentives to change, and they even hold up when clearly preferential economic and material alternatives are offered in exchange for letting go of those values. In fact, using hypothetical case experiments, Atran and his colleagues found that offering subjects large sums of money or other goods in exchange for compromising their beliefs would *increase* the subjects' willingness to commit violent acts in the name of their cause.[18]

So, how do we get anywhere given this mess? At first glance, all this deep evolutionary thought just seems to reinforce our knee-jerk reactions about Middle Eastern conflicts: "Those people will never get along" and "They thrive on conflict." When faced with barriers in nature, organisms either go around them or develop adaptations to live with them, sometimes even changing the nature of the barrier itself in the process. For example, in the tide pools where I work, starfish crawl over rock walls on thousands of hydrostatically controlled tube feet, but their closely related cousins, the sea urchins, actually dig depressions into the same rock walls (resembling the work of an ice cream scoop), transforming them into apartment complexes sheltering hundreds of their fellow urchins. Likewise, when faced with sacred values conflicts, two strategies emerge: we can go around them, or we can transform them.

The idea of going around them requires some counter-intuitive thinking on the part of people who have been banging their heads against the wall of sacred values conflicts for decades. What Atran found in his experiments was that again and again, regardless of age or status, subjects on both sides of a dispute involving sacred values were willing to open their minds to dispute resolution in exchange for materially valueless but symbolically rich concessions from the other side. These "symbolic tradeoffs" often involved nothing more complex than a sincere apology for past wrongs, or a mere acknowledgement that the other group has a right to exist.

When explaining the concept of symbolic tradeoffs, Atran uses an example from a sideshow to U.S. diplomatic efforts of the early 1970s to restore normal relations with China.[19] The deployment of "ping-pong diplomacy" involved sending U.S. ping-pong players to China to compete in highly publicized (within China) matches. Inevitably, the Chinese slaughtered the hapless Americans, scoring a huge symbolic victory for the Chinese while costing the United States almost nothing, because nobody really cared about ping-pong in the United States. That we were willing to concede defeat graciously gave China an alternative view to that of the United States as an unremitting superpower, and helped open the door to ultimately successful diplomacy.

Nelson Mandela also utilized symbolic tradeoffs in shepherding a peaceful transition from apartheid, when he became president of South Africa. In this case—recounted in the 2009 film *Invictus*—black South Africans now in a ruling position moved quickly to change the name and colors of the Springbok rugby team, which had been dominated by white Afrikaners and thus was a symbol of the apartheid regime. Mandela not only overruled this decision—seeing it as a petty form of revenge—but moved to elevate the status of the team and the sport as a symbolic gesture to white South

Africans that their traditional customs would be respected and welcomed as part of a new multiethnic country.

Symbolic tradeoffs are a key—perhaps essential—step in opening the door to negotiation when sacred values collide, but they are unlikely to solve a conflict in its entirety. This is why it is essential to understand that sacred values are themselves not immune to evolutionary pressure. I am not suggesting that the values themselves are handed down genetically, and scientists have not discovered a "sacred values" gene, but analogously to most biological organisms, they are simultaneously extremely well-defined by set characteristics and yet malleable under the right circumstances.

In biology, this apparent contradiction makes sense. An octopus can instantly change its color and skin texture, even its overall shape, to match its environment, yet through all its transformations it remains biologically an octopus. Likewise, it is nearly impossible to eliminate the sacred values held by an individual or group. When they come under threat, they become hardened and more sharply defined. This has clearly occurred with the rise and fall and rise again of right-wing armed militias in the United States. Their high points in terms of membership and high-profile activities coincide with the early years of the Clinton and Obama presidential administrations, both of which were feared to have a strong gun-control agenda.[20] When they are accepted and even left alone, they begin to soften and become more malleable. There is also the possibility of substituting one set of group beliefs for another, provided the right type of "addiction module" is offered, and provided that the timing is right.

Although my focus has been on the negative aspects of belief systems and group membership, they clearly offer benefits to individuals. Like the early bacteria gaining protection with its self–nonself addiction module, belief systems in humans gain us entry into a group that then protects us from others who are not in the group and are potentially dangerous. It is the beneficial aspect of group belief systems that creates an opening by which dangerous beliefs can be substituted with more benign beliefs. A notable example is the twelve-step program of Alcoholics Anonymous (AA), which is one of the most successful addiction treatment programs ever created. Although even adherents don't really know how it works,[21] it is interesting that AA has many of the same characteristics of group belief systems. Participants are asked to take AA's tenets on faith ("let go and let God") and are asked to make costly signals of their loyalty to the group—for example, confessing all of their transgressions to people they've wronged while under the influence—but they also are surrounded by more experienced recovering addicts who protect them from the influences of outside addicts. Here, one more benign addiction module is substituting for a more dangerous form of addiction.

What is the right time to substitute sacred values, or belief systems? In nature the right time can be either when the organism itself is ready for a change or when a change in the environment around the organism makes such change worthwhile. For some hard-shelled organisms like lobsters or crabs, even a minor change, like growth, can only occur when they molt their shells—a period of high vulnerability. Other shelled organisms, such as snails, will radically change their protective layers by growing thicker or growing defensive spikes when they detect the chemical presence of predators in the environment.

For the alcoholic, the timing of the trade between addiction and recovery is almost invariably at a point called "hitting bottom." This could occur when the alcoholic is seventeen or fifty-seven, so it is not so much time-dependent but based on the environment the alcoholic finds himself in. For religious convictions, however, the timing is typically more constrained, specifically to the period around adolescence. Richard Sosis and Candace Alcorta, anthropologists who work at the intersection of biology, culture, and society, have devoted considerable effort to uncovering universal traits of religion, in particular the transmission of religious ideas. They have identified adolescence as a nearly universal period of development when religious ideas and values are imprinted.[22] Moreover, they have shown that there are biological and social aspects of this pivotal period that feed back on themselves and intensify the vulnerability of adolescents. From a basic biological standpoint, the adolescent brain is going through unprecedented changes in growth, structure, and pattern formation—the whole architecture of cognition is changing. From an individual social development perspective, adolescence is a time of risk taking, although how this is related to the physical changes in the brain is not well known. And from a global anthropological standpoint, it is notable that secular and religious rites of passage—which require substantial material sacrifice in exchange for largely symbolic rewards—tend to be concentrated around adolescence. These factors amplify one another, making adolescence both a period with a high natural likelihood of acquiring radical convictions and a period that is actively used by recruiters to indoctrinate group beliefs and values. It is not a coincidence, then, that adolescence is also a prime period for development of drug addiction. The same factors and adolescent dispositions that open an adolescent to addictive states, however, can be exploited to shift a person in this stage away from destructive belief systems. This is why David Dobbs, in reviewing the evolutionary science of the teenage brain, urged us to shift our preconceptions about adolescence, noting, "In scientific terms, teenagers can be a pain in the ass. But they are quite possibly the most fully, crucially adaptive human beings around."[23]

Adolescence is the period where alternative activities may actually have a chance of becoming substitutes for a group identity that compromises security. Here again, an intimate knowledge of the environment and society in which these adolescents live is illuminating. Scott Atran and his colleagues have found that many individual jihadists were recruited as adolescents through local soccer clubs. Sitting as a witness in congressional hearings alongside hardened experts on fighting terrorism, Atran's solutions sound almost quaint. He suggests starting soccer clubs with a secular or at least less radical religious underpinning as a way to provide the same social bonds—even tap into the same ancient need for group identity—but with less extreme side effects.[24] The idea seems logical enough, but developing such alternative pathways to group identity is not free from controversy. A $30 billion crime bill proposed by President Bill Clinton and debated in Congress in 1994 was nearly scuttled in the last minutes by controversy over a relatively minor "midnight basketball" provision, a program designed to offer safe, gang-free spaces for adolescents to play sports at night.[25] Its cost was $50 million, a fraction of one percent of the total expenditure of the crime bill. Atran and colleagues also suggest that new pathways that tap into the "purpose-seeking, risk-taking, adventurous spirit of youth for heroic action"—exactly what adolescents think they are getting when they sign up to attack Western superpowers—must be found and introduced on a peer-to-peer basis, rather than from elders admonishing youth to forge more moderate pathways.[26]

Taken alone, these singular opportunities to alter the course of belief systems appear too rare, and the changes they instill too inconsequential, to alter the enormous evolutionary inertia behind them. It is important to recognize that in most cases, even with very destructive belief systems, a radical change is not needed to be able to live with the risk of any given group. Rather, the change only needs to be just enough so that different individuals or groups can peacefully coexist. Fortunately, such delicate coexistence is as old as the diversity of life on Earth. In fact, a convergence of ideas from early natural history, modern genetics, new views on free market economics, and biological anthropology all point in the same direction—that conflict and cooperation are intimately linked, even dependent upon one another. And in a hopeful sign for security, the history of life seems to show that cooperation serves as the more dominant force. In biology, we call the resultant admixture *symbiosis,* and the remarkable forms it takes in nature and in society are the subjects of the next chapter.

chapter nine

HANG TOGETHER OR HANG SEPARATELY

THERE WERE MANY REASONS that IED attacks in Iraq dropped off suddenly in May 2007. Troops got better at detecting and destroying IEDs before they caused damage. The U.S. government spent billions of dollars on JIEDDO, the Joint IED Defeat Organization, which developed numerous technological solutions, such as jamming devices to disable wireless IEDs. And very late in the game, at the tail end of the decline in IED deaths in Iraq, IED-resistant MRAP vehicles began to lumber onto the streets of Iraq.

But the most important component of reducing IED mortalities in Iraq was *symbiosis*—a working relationship between organisms. Symbiosis between electronic warfare officers from the air force, army soldiers, and marines needed to be forged to develop effective wireless jamming devices. Before that point, electronic warfare experts developed great jamming devices, but when they put them into combat, soldiers and marines discovered that they interfered with their own communication devices, so they shut them off.[1] But as I've argued throughout this book, a technological solution—a thicker shell, a stronger claw, a higher wall, a better jamming device—is rarely a lasting solution. Escalation ensures that a motivated enemy will adapt. JIEDDO officer Noel Lipana acknowledged as much to me, noting that wherever they implemented wireless jamming, the enemy went back to the old wired devices and pressure-plate explosives, and when soldiers went back to cutting wires, the enemy switched to wireless devices. The more important symbioses then were the relationships formed with local leaders, even those that may have once been leading attacks on American soldiers. Like many biological symbioses, these relationships emerged between parties that would seem to have no reason to cooperate. Americans provided technical expertise and resources for civil works projects, and local leaders provided key information on networks of IED makers and their plans. Like many biological symbioses, these relationships were also transformative. When local people saw their leaders working with the Americans, they too began to share information—the peak in tips about IED activity given to soldiers by civilians marked the beginning of a rapid decline in successful IED attacks in Iraq.

So far, I have discussed a number of natural security strategies. In order to survive and change, an organism needs to learn within its own lifetime and across generations. It needs a decentralized organizational system. It needs redundant features. It needs to keep running just to keep up. It needs to reduce uncertainty for itself and create uncertainty for its adversaries. If that organism is a human, it needs to understand human behavior. But an organism or an organization could do all of these things and still easily fail the survival game. That's because of this simple rule of nature: no organism does it alone.

Every organism on Earth is actually a collection of different organisms, bound together by a wide diversity of symbiotic relationships. *Symbiosis* arises in nature for a simple reason. All organisms are constrained in their adaptability at some point, and symbiotic relationships allow them to extend their inherent adaptive capacity to exploit new resources and environments or adapt to their own environment as it changes.

In the biology textbooks, symbiosis falls into three classic divisions. There are *mutualistic* symbioses—relationships in which both parties benefit. There are *commensalisms* in which one party benefits but the other party isn't affected much at all either way. And there are *parasitic* symbiotic relationships in which one party gains and the other suffers. In reality, symbiotic relationships smear across these categories and change with time. Former parasites, like certain viruses that invaded the human genome long ago, have evolved to become harmless or even beneficial. Some mutualistic relationships evolve and turn parasitic, such as once-cooperative relationships between pollinating wasps and flowers that turned parasitic when wasps evolved to lay eggs within the flower (that will hatch into larvae that eat the flower's nectar) without providing the benefit of pollination.[2]

Regardless of the type of symbiosis, three things are certain. First, symbiotic relationships are ubiquitous in nature—you can't go anywhere, look at any biological community, or dissect any biological organism without finding symbiosis. New symbioses are turning up all the time. It's long been known that algae lives symbiotically in salamander eggs, providing nutrients and an oxygen boost to the developing embryo. But recently algae were found to provide the same function *within* the embryonic cells, the first time this has been seen in a vertebrate and something that was thought to be impossible because the vertebrate immune system typically eliminates invaders into embryonic cells. Whole new pathways for symbiosis are being discovered as well. Lars Peter Nielsen has coined the term *electric symbiosis* to describe recently discovered relationships between bacteria deep in ocean sediments and those at the surface of the sediments. The deep bacteria break down hydrogen sulfide to produce energy, but it was thought that they need oxygen from the surface to carry away the electrons resulting from their energy-producing reaction. Apparently, though, nano-scale protein wires form a conductive network between the surface and deep bacteria, and these wires do the job of carrying the electrons to the surface, where abundant oxygen finishes the reaction.[3] In this case, not only are two organisms symbiotic, but they have created a whole physical network to support their symbiosis.

Second, symbiotic relationships are incredibly diverse. There are the obvious ones we see every day, like butterflies, bees, bats, and hummingbirds getting nectar from flowers in exchange for pollinating the plants and maintaining genetic mixing. But some flowers also form a symbiosis with the yeast that sometimes clings to these pollinators, giving the yeast nectar to thrive on and getting warmth from the metabolic activity of the yeast, which in turn allows the flower to release more of the chemicals that attract pollinators in the first place.[4] There is no one pathway toward successful symbioses between species.

Third, symbiosis creates reactions that are more than just the sum of two organisms working together. Symbiosis creates *emergent* properties that you wouldn't predict from just looking at the two organisms on their own. That is to say, symbiosis *transforms* an organism and transforms the environment around the organism. The relationship creates whole networks of interactions, builds new habitats for other species to use, and even changes the tenor of conflict in the larger ecosystem. Lichens, which grow proficiently on bare rock and even in icy valleys of Antarctica, are actually a symbiotic partnership of fungi, algae, and bacteria. Through the process of their cooperative growth, they physically break down bare rock into usable soil that completely unrelated plants and animals can thrive in. The lichen's contribution is not trivial—

it's estimated that the biomass of lichens on Earth exceeds the entire biomass of the world's oceans.[5] Deep-sea fish ensconce light-emitting, or bioluminescent, bacteria within special organs on their skin, so that they are able to hunt and lure prey in the inky black depths. In an amazing show of "convergent evolution," symbiotic bioluminescence has evolved many times independently. More precisely, the ability to glow may have evolved only once in bacteria, but many different fish, squids, and other invertebrates have independently evolved their own structures to house these valuable bacteria,[6] an amazing feat of transformation, facilitated by a tiny bacterium.

Other transformations are grotesque—some parasites actually force their host to change its behavior to further serve the parasite. There are worms that change the behavior and appearance of their ant hosts so the ants resemble fruits favored by birds that serve as the final host of the parasite. Parasites on small aquatic crustaceans called amphipods cause them to abandon their normal behavior of hiding under rocks and instead swim to the surface, where they are ingested by birds that serve as a definitive host within which the parasite can reproduce.[7] Humans don't escape this manipulation. Guinea worms—the treatment of which I will discuss below as one of the great success stories of human social symbiosis—are ingested by humans who drink water with guinea worm–infected fleas. Once inside the body, they cause intense burning pain in their human hosts, which causes people to seek out water to soak their skin, allowing the worm to release its larvae, which can then reinfect the water that another potential human host will drink.[8]

But the behavioral changes created by symbiosis may have wider benefits as well. Studies on monkeys and apes show that when individuals are forced to begin a cooperative relationship (to help one another get food, for example), conflict overall between the animals is reduced.[9] Large fish on coral reefs that develop symbiotic relationships with cleaner wrasses—small fish that set up "cleaning stations" to eat parasites out of the larger fish's mouth and gills—are less aggressive not only to their cleaning partners but toward all other fish on the reef as well.[10]

The early-twentieth-century ecologist Warder Allee was a lifelong proponent of the transformative capacity of relationships among natural organisms. Over decades of natural history observations and experiments in his lab, he studied two major aspects of the social relationships of animals. First, he studied the supposedly cruel part of biology—the fierce competition for resources, the pecking orders, and dominance hierarchies that pitted one member of a group against another. But he also studied, and one suspects greatly favored, the cooperative side of nature, the deliberately altruistic and instinctual connections to other living beings, what he called "automatic" cooperation between animals of the same and different species.

Where Allee did look at tooth-and-claw competition, he generally saw it as a stabilizing force. But his perspective wasn't that of a Victorian Social Darwinist; in his view, it wasn't the sole "alpha" individual, through its superior breeding and intellect, that rightfully kept the other inferior subjects properly in line. Rather, he saw that the *interactions* between organisms in a pecking order were essential to the group's overall stability. In one test of this idea he observed pecking orders in two hen houses. In one, the chickens were allowed to develop a normal dominance hierarchy, and Allee observed that this stayed stable for months at a time, maintained by constant small reminders—little pecks—between more and less dominant individuals. In the other house, the hen that emerged as dominant was always removed after a few days, throwing the pecking order into chaos. In this house, there was far more violent fighting as power vacuums continually opened up.

But it was cooperation that enthralled Allee. He collected dozens and dozens of examples of cooperation through his own painstaking laboratory experiments, field observations, and reviews of decades of biological research. He found embryos that cleaved faster when fertilized in groups. Protozoans, which reproduce through cleavage, produced much greater numbers when two were put together in a dish than the combined numbers of clones produced by two individuals housed separately. Goldfish in groups survived longer when exposed to toxic silver than they did as individuals exposed to a proportional amount of the same toxin. Male manakin birds in Panama formed lines in forest clearings, singing to attract females. Although they were competing for mates, working in a group was much more effective in attracting females than if they sang alone in isolated patches.

Allee worked to reconcile these two apparently contradictory biological forces—competition and cooperation—not just in the strictly biological realm, but as they related to the social affairs of humans. He reminded readers that Adam Smith, whose *The Wealth of Nations* is practically a Bible for libertarians and free-market conservatives railing against government regulation, also wrote *The Theory of Moral Sentiments,* which suggests that sympathy for fellow humans is an essential force that counteracts the self-interested nature of humans.[11] And from Allee's hen-house experiments he extrapolated, "Person to person competition, if not too severe, may lead to group organization which increases the effectiveness of the group as a cooperating social unit in competition or cooperation with other social organizations."[12]

This statement suggests the kind of nested, recursive process that we see all over biological systems. That is, if competition within a group creates group cooperation that can be used in the group's competitions with other groups, then those higher-level competitions should lead to cooperation between the groups that can be then used at yet higher-level competition. In this view, cooperation, like learning, is its own evolutionary force that contributes to an organism's immediate survival but also creates the possibility for adaptive responses to future challenges.

In the end, although Allee felt that both the "egoistic" and cooperative forces were essential for evolution, he had much more belief in cooperation than the brutish stability of dominance hierarchies as a model for human affairs. He argued that each more complex level of biological organization, as well as each more complex organism in evolutionary history, was built on the foundation of cooperative arrangements at the levels below or before them.[13] For example, multicellular organisms could never exist if some form of lasting cooperation did not occur between their single-celled ancestors.

Allee's bias toward cooperation is owed in part to his understanding that all biological and social systems change, and he saw cooperative relationships, rather than dominance hierarchies, as adaptable to change. That he eagerly brought this bias to his speculations about human social relations likely stems from his personality and the time period in which he lived—he was an optimist living in a time of global upheaval. His work and life were bracketed by two world wars and were witness to the collapse of longstanding empires and the relatively quick termination of newer empires in Germany and Japan, brought about by extremely costly but effective cooperation among nations. Unlike the typical Social Darwinist, Allee saw biological tendencies as pointing decidedly away from warfare:

> Such evidence and reasoning as I have presented indicates pretty clearly that the present system of international relations [based on war] is biologically unsound. Attempts that have been made in the past to lend biological respectability to

> the existing system by regarding it as an expression of an inevitable struggle for existence have overlooked not only its defects as a selecting agent, but more serious, have often not even been conscious of the existence of another fundamental biological principle, that of cooperation.[14]

With this worldview, Allee saw the postwar era as an opportune time to create a cooperative organization among nations, to ensure a lasting peace. While this was a common sentiment of the era, and ultimately led to the United Nations, his perspective was unique in that he felt that the design of this organization should be biologically based, noting that "the biologist's international system must be a dynamic organization capable of and designed to effect changes rather than set up to preserve any given status quo, regardless of how favorable for the predominant powers."[15]

In particular, Allee defended the creation of such a large organization by insisting that it take on the same nested and recursive processes seen in nature: "The maintenance of smaller cooperative and competing units within the larger one is part of the scheme as sketched."[16] This type of organization mirrors the decentralized adaptable organizations I discussed in Chapter 4, but is sadly quite different than the international bureaucracies—the United Nations, the International Monetary Fund, and the World Bank—that ultimately came into being after World War II.

Jean-François Rischard, a former World Bank president, has recognized the failure of these organizations to be adaptable to pressing world needs. His solution is to facilitate the creation of "global issues networks"—essentially bottom-up nongovernmental partnerships of individuals and groups that are borne out of local motivation to find solutions to specific problems.[17] Rischard's vision, which provides a pathway to create cooperation among entities that are currently in conflict, such as government and industry, is essentially an updating of Allee's "biological" vision for a global peace-keeping entity.

Allee's ideas fell out of favor soon after his death in 1955, which coincided roughly with the discovery of the structure of DNA and the rise of molecular biology. I think the two events are related. Watson and Crick's explanation of the elegant structure of the molecule of life set off a revolution in biology, creating a chain reaction of amazing discoveries about life that have continued to emerge to this day. But the direction this revolution led was decidedly away from the natural history–based observations of Allee and even more so away from his ideas about broad-scale cooperation. Molecular biology was all about incredible things happening at very small scales. DNA imprinted *individuals,* not groups, and when natural selection happened it happened to individuals and to genes themselves, certainly not to groups. And if his notions about cooperation and group behaviors seemed antiquated in the new, more precise world of molecules, Allee's broad speculations about what the pecking orders of chickens say about the affairs of man seemed completely unglued, the ramblings of a doddering old naturalist in an era where answers came by focusing ever sharper at ever smaller scales of life.

But now Allee's ideas are coming back into favor. It started a few decades ago as biologists became alarmed at the global decline of many populations of plants and animals.[18] We began to refer to "Allee effects" to talk about the minimum critical mass of animals needed to keep a population going. For example, as abalone are depopulated on the West Coast due to overfishing and disease, they may simply not live in dense enough aggregations for their free-floating eggs and sperm to meet in the turbulent Pacific waters, and thus their decline is accelerated. This attribution alone would be a nice footnote to Allee's contributions, at least worthy of the short Wikipedia entry that bears his name.[19] But the more fundamental, and I'd argue more important, contribution from Allee—that cooperation is a fundamental biological force—is also coming back to the fore, this time repackaged in the more inclusive concept of *symbiosis.*

THE UBIQUITY OF SYMBIOSIS

The transformative power of symbiosis is best illustrated in the concept of symbiogenesis, the idea that many integral parts of organisms today were once independent free living organisms themselves. In a symbiogenetic view, the world trends toward ever more cooperation, even as it grows more complex. Ironically, while it may have been genetics that stalled serious further exploration of Allee's ideas, it was a hardened geneticist (albeit one with a nonconformist streak) who helped bring it back.

Lynn Margulis, who is now a well-respected member of the National Academy of Sciences and credited for her work advancing symbiogenetic theory, was once a scrappy student who advanced herself to upper grades in Chicago public schools under not entirely forthright premises and graduated from the University of Chicago at age nineteen.[20] It's a good thing that Margulis was both brilliant and a nonconformist, because she began her career in genetics by focusing not on the cell nucleus where DNA is concentrated (and where nearly all her colleagues were focused), but on the oddball genes hiding out in organelles like mitochondria, in the soupy cell cytoplasm outside of the nucleus. By looking where most other scientists were not, she was among the first to identify likely targets of symbiogenesis and develop a credible theory of how symbiogenesis could come to be. Like her public school records, her ideas were once considered fanciful fabrications. But just as she proved to her Chicago public school that she was more than competent at her adopted grade level, her work ultimately bore out the validity, indeed the ubiquity, of symbiotic origins of complex life forms and essential life processes, like photosynthesis and energy conversion. Where this idea merges with Allee's earlier view of cooperation is that Margulis sees symbiosis as arising out of conflict that transitioned into cooperation. In her words, symbiosis is "a palpable legacy of a violent, competitive and truce-forming past."[21]

This creative tension between competition and cooperation expressed in symbiosis is an intriguing window into human conflict, which is also a product of competitive and cooperative elements. We tend to focus on the obvious—that competition, even violent competition, is widespread across human societies. It is found between individuals, families, villages, cities, schools, sports teams, religions, and nations. But so too is cooperation widespread in human societies, as it is among social animals such as bees and wolves.

The roots of competition seem fairly obvious, but there is still spirited debate over how cooperation arises. Why would we want to help another individual, especially when there is likely to be a cost to us for doing so? In some views, cooperation arises from strict cost-benefit analysis. The benefit to my reproductive fitness—the likelihood of passing my genes to my offspring—or even just being acknowledged for my cooperative spirit may outweigh the cost of not being selfish. If a potential mate finds me more attractive because I help an old lady across the street, that is certainly worth the cost of time wasted.

Alternatively, this cost-benefit may not be deviously figured in material terms but rather calculated subconsciously in the currency of genetic relatedness. This is the concept of "kin selection," in which there is a measurable benefit of helping and protecting not just yourself but those who are closely related to you genetically. This is why workers of some bee colonies (who are all sisters) will protect the hive to their deaths—if they can help their closely related sisters live, it may be worth sacrificing their own life. But kin selection isn't a satisfactory explanation of cooperation in humans.

We help fellow humans who are only distantly genetically related, and we even help other species who are more distantly related, sometimes sacrificing our own lives in the process.

Cooperation may also arise through the expectation of reciprocity, so that every apparently altruistic act comes with an expectation of a return. If a chimp picks the fleas off another chimp and the other returns the favor later, they are more likely to cooperate in the future.[22]

On the flip side, cooperation may also arise due to fear—there could be social ostracism or more serious punishments for those who don't cooperate. Workers in some insect societies that don't work but instead try to lay their own eggs are punished by the queen, who might kick the cheater out of the nest, or by fellow workers, who may destroy eggs not laid by the queen.[23]

Finally, echoing both Allee and Margulis, primatologists and ecologists have argued recently that competition between humans led in a recursive fashion to cooperation. Specifically, in this theory human cooperation arose through selection for more cooperative individuals. Groups in conflicts that had more of these individuals did better in conflicts, gained more resources and access to reproductive females, and therefore mated more and enriched—through some combination of genetic dispositions toward cooperation and societal reinforcement of cooperation—the population in cooperative individuals.[24] Theoretical models based on social network theory have also shown how cooperation at larger scales can arise from small-scale competitions and conflicts among individuals.[25] In modeling terms, cooperation acts as an "attractor" to which the whole system of interacting individuals settles into under most realistic simulations.

These theoretical musings on how we became a cooperative species are supported by the indelible imprints of past cooperative behavior on modern humans. There is experimental evidence that centers of the brain associated with rewards get active when humans enforce behavior that increases group cohesion,[26] and studies on twins reveal that at least some of our altruistic disposition toward promoting fairness is genetically inherited.[27]

Clearly, human scientists can use their own complex minds to come up with all sorts of theories about human cooperation. They have even devised experimental tests of cooperation in the realm of game theory, where human research subjects are given a hypothetical situation in which they must make a decision that will have costs and benefits for themselves and for other people playing the game. They might have to decide whether to turn in a fellow "criminal" or remain silent and receive a bigger punishment, or they might be asked how much of a pool of prize money to give to the other subject under the condition that if the other subject doesn't agree to the division, both parties will get nothing.[28]

These simplistic sounding games, which are designed to study incredibly complex human behaviors, are simple for a reason. They attempt to peel back layers of societal veneer and idiosyncratic individual experiences by putting people in a very sparse world without economic inequalities, political parties, and religions, to get at the unadulterated roots of human cooperative behavior. Despite their simplicity, they often reveal fascinating and sometimes unexpected behaviors. Nonetheless, taken together over decades of game research, they reveal something we already know well: humans are complicated. There is no one answer about cooperation that comes out of game theory, but rather a wide range of human behavior gets revealed as the rules of the game are changed by different researchers. People may be less inclined to cooperate in one-time games but become highly cooperative when there is the possibility for the other side to retaliate for a punitive decision. People playing in teams make different decisions than people playing alone. People playing face-to-face games cooperate more than those playing over computer, and people who are just shown the image of a relative or merely a picture of a human eye cooperate more than those who are given no hints that they are playing against a fellow human being.[29]

None of the extensive research so far unequivocally answers the question of whether there are truly selfless acts in nature or just an endless series of I.O.U.s. In the end, the question is less relevant than the fact that cooperation unequivocally exists in nature, that it is not only compatible with competition but perhaps dependent on competition, and that cooperation takes a lot of different and sometimes surprising forms.

THE MOST UNLIKELY FORMS OF COOPERATIVE SYMBIOSIS

What all this does tell us about survival in society is that we have vastly underutilized the lessons of symbiosis and cooperation. This is important because when we look at security situations in society, we tend to be overwhelmed by seemingly intractable conflicts—Israelis versus Palestinians, radical Muslims versus the West, Red Sox versus Yankees fans—where cooperation seems impossible.

But nature reveals the power of symbiosis best when we see it occur between species that at first glance would seem to have no business at all in cooperating, such as large predatory fish and the much smaller fish that clean their teeth or hitch a ride on them. In society, as in nature, mutualistic symbiotic partnerships between the most unlikely of collaborators are developing and ameliorating potential security crises around the globe. My colleague Terence Taylor, for example, has helped incubate symbiotic partnerships between Israelis, Palestinians, and Jordanians,[30] as well as practitioners from six traditionally hostile countries on the

Mekong River, all working together to identify and neutralize disease outbreaks on whatever side of borders they occur.

Several features of these cooperative networks should be recognized. First, the networks have demonstrated success even beyond the feat of getting members of mutually hostile nations to work with one another. Network practitioners were quietly allowed into notoriously restricted Myanmar to do their work days, not weeks, after the catastrophic cyclone there. The Middle East consortium was ramping up responses to H1N1 "swine flu" days before the World Health Organization began to increase its alarm.[31] They have also resulted in professional benefits for their participants. Members of the Middle East consortium collaborated on the first scientific paper jointly authored by an Israeli and Palestinian since World War II.[32] Their success, like a properly functioning evolutionary feedback cycle, has encouraged further success. Large corporations such as IBM have been impressed by these networks and have contributed vital database technology. Better still, new consortia are being replicated, for example in southern Africa, based on the successful performance of the original Middle East consortium.

Second, these networks weren't mandated by high levels of government or through international treaties but have emerged from the ground up as local, adaptive responses to a real need to protect regional food supplies and human health from pathogens that know no borders. Health practitioners and health ministers, not prime ministers and heads of UN programs, envisioned, created, and continue to operate these networks. They are, in fact, excellent examples of how decentralized organizations can be more effective and adaptable than better-resourced and more powerful centralized governments. Indeed, while governments often put up metaphorical and real walls between one another, decentralized organizations help facilitate mutualistic symbiosis.

Third, the networks were not designed to tackle the much larger and complex issues of creating peace between their member states, though they very well may be an opening to further peace agreements. Symbioses in nature never solve all of an organism's security problems, but where new symbioses develop, they arise to solve the most immediate challenges. It may be that later, these relatively limited acts of symbiosis build into amazing outcomes that no one could have predicted.

Finally, the networks call to mind the necessity of symbiosis. In nature there are "facultative" symbioses, which provide certain benefits to organisms but are not absolutely necessary for the survival of one or both of the organisms. The symbiotic zooxanthellae algae that give sea anemones their bright colors are a good example. When placed in the shade under a dock or in an aquarium tank, such as a sorry looking West Coast anemone I saw in the New England Aquarium once, the sunlight-using algae leave or are expelled from their anemone host, which continues to live, pale and bleached, but otherwise just fine, without the algae. But there are also "obligate" symbioses, where the organisms will die without their symbiotic partner. These are especially apparent in harsh climates. When biologists first explored the deepest parts of the oceans, far from the sunlight that jump-starts nearly all biological interactions on Earth, they were awed to find rich thriving ecological communities built around spectacular colonies of large tubeworms. The worms had no mouths or guts but were well populated by obligate symbiotic bacteria that fix hydrogen sulfide into food for them. As the consortia in the Middle East and the Mekong continue to work together, they are rapidly evolving to become obligate symbionts. The networks greatly expand the capacity of any individual member state, giving them a built-in impetus to continue—without the network, each individual state would not only be powerless over outbreaks in neighboring states but would also be much less capable of tackling diseases within its own borders. Thus, they create a web of interdependence, like the bacteria in ocean sediments and the cleaner wrasses on the coral reef—and this in turn transforms a small effort to cross physical and ideological borders into a self-sustaining and mutually beneficial partnership.

A recent updating of the "mutually assured destruction" doctrine for an era of asymmetric threats incorporates the idea of obligate symbiosis. In this approach, called "mutually assured support," alliances between nations are predicated on the assurance of support—through resources, public signals of support, and punishment of defectors—in the case of attack by any nation or regime against one of the alliance members.[33] In this way, members of the alliance are assured a better capability to survive and respond to an attack than they could on their own. This symbiosis is mutually reinforcing because the benefits would only accrue to nations or regimes that did not attack other alliance members.

Cooperation does not always arise out of conflict, but when it does, it is often in cases where a biological need trumps ideological differences behind the conflict. Sometimes these can be fairly mundane—German and British troops in World War I occasionally held unofficial cease-fires to retrieve wounded and dead soldiers in the no-man's land between trenches.[34] Like the disease surveillance networks, these cooperative agreements were not sanctioned by the soldiers' respective governments and were looked down upon by the commanding officers; rather, they were self-organized between enlisted men from the respective warring sides.

Conflict-borne cooperation can stretch from a single front to an entire war. In 1995 former president Jimmy Carter helped negotiate a two-month cease-fire in the ongoing civil war in the

Sudan so that medical teams treating guinea worm outbreaks could do their work.[35] This window of opportunity was essential in getting a foothold on what was a growing problem in Africa. Now Guinea worm treatment has so greatly reduced the caseload in Africa that it may soon become one of the few public health examples of successful eradication.

Symbiosis is a powerful tool, but it is not infallible, in part because the benefits are so context dependent. Hydroids living on hermit crab shells are usually beneficial because they keep off other things that might foul and damage the shell. However, in some areas, a parasitic worm seems to prefer hermit crab shells with hydroids, and the worm weakens the crab's shell enough to make it vulnerable to crushing predation from blue crabs.[36] So, a relationship that works in one area may not work in another. Symbiotic relationships between local leaders and U.S. forces worked extremely well in Iraq but have had at best a mixed record in Afghanistan, where U.S. forces worked to create similar symbiotic partnerships even before IEDs became a major threat. In some cases local leaders made symbiotic partnerships with *both* U.S. forces and their enemies, and IED attacks against U.S. forces steadily increased from nearly nonexistent to become responsible for almost half of all combat deaths there.

Although Allee and Margulis saw the creative and transformative nature of cooperation, there is a more dour utilitarian view of mutualism that dominates among experimental biologists. They note that mutualism gives rise to cheaters; that mutualism is not always cooperative, with one side getting much more out of the relationship than the other; that mutualisms have hidden costs and must be maintained by brute coercion. There is certainly evidence to support this line of thinking, from the animal kingdom to kingdoms of humans. For instance, a textbook case of mutualism is the *Rhizobium* bacteria that live in the roots of legume plants such as alfalfa and soybeans. The bacteria "fix" nitrogen into a form that the plant can use nutritionally, and in exchange the plant provides a safe home and oxygen from photosynthesis for the bacteria. But a clever experiment, in which nitrogen was replaced with argon—which allowed the bacteria to live but did not provide fixed nitrogen to the host plant—revealed that the plant will punish the bacteria by withholding oxygen if the bacteria fail to give the plant its fix of fixed nitrogen.[37]

Yet the view that mutualism is just a market transaction may belie accounting errors on the balance sheet of benefits and costs to the organism in a mutualistic relationship. Like a management consultant swooping in to analyze and "right size" a company, or a war correspondent embedded for a couple of weeks in a decade-long war, the biologist taking a few months or even years to study a mutualism that has been going on for millennia may miss key benefits that accrue through sustained cooperative relationships.

Warder Allee spent much of his life trying to discover the new thing that appeared when species or individuals cooperate, that thing by which species in groups did better than organisms alone. For example, in the goldfish exposed to toxins, that thing was found to be a slimy chemical that the fish only secreted in groups, which seemed to buffer the toxic effect. These little mechanisms cannot be predicted by looking at solitary individuals, nor can they in a sense be accounted for by merely summing the balance sheets of each individual in a mutualism. They are what researchers of complex systems call *emergent properties*. Allee never discovered all the emergent properties supporting the cooperative relationships he studied, but contemporary scientists are finding surprising emergent properties wherever they look at symbiotic systems—the peace that reigns on a reef stationed by cleaner fish, the rich soil provided by the algal-fungal-bacterial threesome that makes up lichens, and the joint publication of epidemiological research by Israelis and Palestinians.

Even the biologists who take the "free market" approach to mutualism and cooperation acknowledge they have an unanswered conundrum.[38] If mutualism is just about using another being to maximize your economic return, it should *not* be evolutionarily stable—it's a race to the bottom for which one partner can extract more and give less to the other partner until there's no reason at all to cooperate. Indeed, biologists have invoked Garret Hardin's famous Tragedy of the Commons construct to describe the ability for one party to exploit mutualisms. In Hardin's view, common property, such as grazing land on an old English "commons," or fish in the sea, is subject to overexploitation if everyone acts in their own self-interest and takes what they can from the common resource without regard for others.

So the conundrum is, if mutualism is so unstable, why is it so common in nature? Darwin took painstaking care to formulate his argument that unstable, intermediate evolutionary forms don't stick around long enough to make an impression on the fossil record that we can find today. In fact, this was one of the most important arguments in the development of his theory of natural selection. But mutualisms past and present are abundantly available for study, not something you'd expect from an unstable relationship.

Elinor Ostrom, who won the Nobel Memorial Prize in Economic Sciences in 2009 for her work on how people in society solve common property exploitation problems, may be informative for figuring out why cooperation in nature, and in human affairs, is more common than mere free-market accounting would suggest. In Ostrom's more subtle accounting, humans can create all sorts of formal and informal rules that keep them from overexploiting common property. These include ways of controlling access to the property. A fishing cooperative I've worked with in Baja California, Mexico, does this by only giving the oldest men in the village access to fish abalone—the highest value and easiest-to-catch seafood that is also the most in danger of being overexploited. Younger members of the co-op have access to successively less valuable resources.

In Ostrom's view, using graduated systems of punishment, starting with largely symbolic gestures, is another way to induce cooperation. In a Mexican fishing village in the Gulf of California that has managed to control overexploitation of resources, fishermen who are caught poaching face serially increasing punishments, starting with the embarrassment of having their fiberglass *panga* boat taken during the night and put up on cinderblocks in the middle of town.[39]

The fact that contemporary researchers with multi-million-dollar molecular labs, Nobel Prizes, and appointments to elite scientific societies have discovered actions and rules in nature and society that turn conflict into cooperation is important to our study of security, but it should not eclipse the contributions of an old natural historian who used little more than his keen observational sense to confirm his belief that cooperation is the strongest organizing force in the social lives of all animals, including humans. As this naturalist, Warder Allee, wrote, "We should not overlook the existence of strong, competitive, egoistic drives among all animals, ourselves included. These must be duly considered in any workable plan for a world order. Our job is to keep them in their true place, somewhat subservient to the even more fundamental cooperative, altruistic forces of human nature. They should not again be allowed to steal the international show."[40] In other words, moving beyond conflict may be in part facilitating pathways for our natural cooperative tendencies to take their course.

Here we come to a conundrum. Throughout the book, I've discussed a number of different powerful ways—from decentralized organization to symbiosis—that organisms can become adaptable to change. But we well know that forces of nature can overwhelm almost any system, no matter how adaptable. The dinosaurs spent far more time on Earth than we humans have, but they were wiped out in a virtual instant when a large impact caused fires and scattered ash and dust across the globe, causing a rapid global cooling that cold-blooded reptiles couldn't cope with. These overwhelming forces of nature, expressed in rare and unexpected events, are what make the evening news headlines: "Mother nature takes her revenge!" "Nature's fury unleashed!" and the like. But that power of nature can also keep us secure. It often comes down to fighting fire with fire, or, more precisely, fighting microbes with microbes, and fighting tropical storms with wetlands. The next chapter focuses on the existing systems of immunity and ecology that keep us secure when we don't do a thing to help them—*especially* when we don't do a thing to help them. Their power is readily available to us, usually free of charge, provided we don't inadvertently weaken it as we rush to create our own "intelligently designed" security systems.

chapter ten

WE HAVE MET THE ENEMY AND HE IS US!

We are well aware of how individual components of the Earth's complex biosphere can play a role in security. Hannibal invading Italy with an army of elephants. Lawrence of Arabia leading guerilla fighters on camelback. Apache raids on war horses. Carrier pigeons delivering top-secret messages across occupied Europe. Bomb-sniffing dogs. Drug-sniffing dogs. Even cancer-sniffing dogs. Dolphins finding submerged mines in harbors. Chimpanzees testing space capsules. Mice testing antidepressants. Chicken eggs that incubate viral vaccines that prime our immune systems for coming epidemics. The direct use of nature to ensure our security, usually through trained or confined organisms, is rooted in our legends and history, our technological triumphs, our greatest successes in social medicine, and discoveries both vitally important to our survival and highly controversial in their methodologies. In every case, nature helped us stretch beyond our own abilities, providing a cheaper, easier, and more reliable way to do things we couldn't or wouldn't do on our own.

Despite the profound influence these organisms have had on our survival, the use of nature in this way has become so commonplace that it almost fades into the background, hardly worth calling out for special attention in a discussion of innovative ways to use nature in security. Or perhaps, because we imposed our own clever modification on these organisms, we've even forgotten their identities as natural organisms, considering them merely as tools, not essentially different from an MRAP or a crash-test dummy or any other pill from the pharmacy. And yet these natural tools, largely taken for granted, only represent a small fraction of the ways nature provides security for us every day. Our most important natural security systems are those that haven't been deliberately changed by us at all. They are just there, doing their natural thing and keeping us safe.

This chapter focuses not on nature-*inspired* security systems, but on natural security systems that have developed wholly in the setting of biological evolution. These are a subset of what has been called "nature's services"—things of high value to humans that are provided free of charge by nature. In keeping with a key theme of this book, they can be found at every level of biological organization, including deep within our own bodies. Some of these services have been long forgotten, some are just being discovered, and some have been recently revalued at a much higher rate following tragedies that struck in their absence. All of those that we have some control over, I argue, are still vastly undervalued and underutilized. In our haste to let experts design and then deploy security systems, we often ride roughshod over these truly natural security systems, irrevocably damaging them. Even worse, we often modify these systems in the name of greater security, and our intelligently designed security systems usually turn out to do worse on the balance than nature in its unmodified state.

Take our feet. They are a wonderful security system that keeps us balanced, allows us to move fast and are almost always our first responders to the changing environment we move through—and we've been busy breaking down and replacing them with our modern (and inferior) technology. Christopher McDougall's book *Born to Run,* which is part ethnography of the long-distance running Tarahumara Indians of the Copper Canyon in Mexico and part scathing indictment of the running shoe industry, makes the case that our feet have always been the first and best way our species has interacted with the world. Our emergence from the rest of the primate line co-occurred with a body well-adopted not just for an upright gait but adapted like no other animal for running long distances. McDougall argues that the Tarahumara can run nearly endless distances in some of the roughest terrain in the world without injury—well into old age, by the way—because of, not despite, the fact that they run barefoot or in the thinnest rubber sandals. By contrast, most of us have bought into the idea that we need to protect our arches by supporting them (but have you ever seen an architectural arch with support *under* it?), massively pad our heels, and then (probably because our heels and arches are now unnaturally high off the ground) augment this protection with all sorts of "stability" devices, springs, air pouches, gel inserts, torsion control bars, and the like. To "safely" go running, we've got to drop $100 or so for something we used to be able to do for free. What's worse, according to McDougall and a growing number of peer-reviewed studies, these shoes aren't protecting us from all our knee and hip and Achilles tendon and plantar fasciitis injuries, they're *causing* them.

So, how can our bare feet protect us? The same way as other natural security systems—with a lot of redundancy and remote sensing. McDougall's description of the foot's architecture nicely blueprints the redundancy: "Buttressing the foot's arch from all sides is a high-tensile web of twenty-six bones, thirty-three joints, twelve rubbery tendons, and eight muscles, all stretching and flexing like an earthquake-resistant suspension bridge."[1] This redundancy, when allowed to work, does a great job of shock absorption, but that only explains part of why barefoot running works. Instead of accepting one centralized direction to pound down and through the heel and maintain that exact heel strike no matter what the terrain (as a super-cushioned running shoe was intentionally designed to make the leg and foot do) a bare foot lets all those redundant muscles and tendons and joints and bones (not to mention the many sensitive nerves of the sole) take over as semi-independent agents sensing the ground. Like Geerat Vermeij "seeing" shells with his fingertips, barefoot runners start to "see" the complex terrain with all the sensory equipment of their feet. McDougall recalls a Nike researcher filming barefoot runners and coming to the ironic conclusion that running barefoot was a lot better than running in Nikes: "He was startled by what he found: instead of each foot chomping down as it would in a shoe, it behaved like an animal with a mind of its own—stretching, grasping, seeking the ground with splayed toes, gliding in for a landing like a lake-bound swan."[2]

Of course, it's certainly up to individuals to determine how much they want to trust their own body versus how much they want to trust Nike and Doctor in a Box. I will say that after struggling through early middle age with a growing number of running injuries, I decided to take a leap and try running barefoot. Almost instantly, all of those nagging knee pains, Achilles strains, and a bad recurring case of plantar fasciitis disappeared and haven't plagued me for two years, so on a personal level I'm a convert. But individual success using natural security systems doesn't automatically scale up to institutional conversion. Public health, for example, could benefit enormously from utilizing natural services but has been slow to embrace them.

NATURAL SECURITY AND MEDICINE

Our bodies have all sorts of other natural defense systems that we often foolishly and deliberately override. These defense systems were developed over hundreds of thousands of years of human evolution, and many can be traced back to organisms long predating humans in evolutionary development, but we dismiss this acquired wisdom in trying to quickly alleviate the temporary discomfort that often accompanies our defensive responses.

I can remember my daughter's first fever; not extreme, but it lasted a while and kept her from sleeping, and being a concerned and sleep-deprived first-time parent, I brought her to the Doctor in a Box clinic. The doctor on call took a quick look at her, told me it was nothing serious, and prescribed that I give her some Tylenol to stop the fever. Since it was a slow night at the clinic, I decided to press the issue a bit.

"Doc," I asked, "I don't know much about medicine, but isn't a fever a sign of the immune system fighting off infection?"

"Basically, yes," he replied.

"So, then might it be better to let it do its job than suppress it with drugs?"

"Well," he demurred, "it probably is, for a fever that's not too high like your daughter's, but working here I just usually want to get the kids and their parents comfortable as soon as possible, so I tell them to kill the fever."

Indeed, both of us were right. Fevers are uncomfortable for patients and for their caregivers, but they are also a natural security system that's been tested over a long time. All our vertebrate cousins, from fish to birds, get fevers, and newer research shows that the vertebrate immune system functions better when it's hot.[3] The fever is essentially prepping the operating room for the immune system to get to work excising an invading pathogen.

This disconnect between the deeper knowledge of medicine—that is, getting at the roots of why we get sick—and the shallow practice of medicine—basically treating symptoms—is what drives the field of "Darwinian medicine." Darwinian or evolutionary medicine strives to identify the evolutionary strategies of disease and aims to develop a method of practice that respects the power behind our natural responses to disease. For example, conventional doctors are trained to understand that the proximate cause of jaundice is a buildup of bilirubin, so they work to get rid of it. But bilirubin isn't just a waste product—it has evolved to function in humans as a powerful antioxidant that slows the ageing process. A more critical example is the over-prescription and over-the-counter use of antibiotics. Evolutionary biologists understand that everything from bacteria to insect pests will, through selection for the hardiest varieties, acquire resistance to broad spectrum attempts to eliminate them. The consequences of this lesson of evolution are beginning to be appreciated by mainstream health practice circles. The rise of super-resistant bacteria, especially in places like hospitals, where antibiotics are highly concentrated, is now recognized as an enormous threat claiming a large proportion of the estimated 99,000 annual infection deaths in U.S. hospitals alone.[4] Nonetheless, as I discussed earlier, incorporating evolutionary ideas into other fields of practice has a mixed track record. Although the core ideas of Darwinian medicine have been around for several decades, articulated most clearly in Randolph Nesse and George William's book *Why We Get Sick,*[5] evolutionary biology is still almost completely absent from medical school curricula.[6]

The specter of emerging infectious disease is of grave concern to public health officials. New diseases for which we have little natural immunity are just a few mutations away from being safely contained within animal hosts to something that can infect a huge and available new human population. Already our best defense against these diseases is almost purely natural—basically identifying their character and developing vaccines that are still often cultured in the albumen of chicken eggs (which provide a nice living substrate for growth without having to sacrifice a lot of animals to make large batches).

But a complementary line of defense would be to lower the chance of infection itself by lowering the prevalence and virulence of diseases. Here nature may provide a very simple solution—just keep the variability of nature intact. At least in cases of some diseases, like Lyme disease and hantavirus, which utilize multiple hosts to spread, having a diverse array of potential hosts around can weaken the strength of the disease overall.[7] This is because not all species are equally good at transmitting the disease—for example, 90 percent of mice bitten by ticks transmit Lyme disease, whereas only 15 percent of squirrels bitten do so[8]—in essence, the total population of the disease gets diluted by passing through a higher proportion of ineffective hosts. Similarly, West Nile virus, which has been gaining a foothold in the United States, increases its prevalence when bird diversity declines. Diseases such as schistosomiasis, which uses aquatic snails to incubate the infective larval form, are reduced in the presence of a healthy population of predatory fish that eat the intermediate snail hosts.

In general, it follows that protecting the habitats that protect all this diversity is likely to help control disease. Scientists have recently found that clearing forests greatly benefits the mosquitoes that spread malaria.[9] Where key habitats, such as the wetlands around New Orleans, Louisiana, are filled or converted for human use, prevalence of West Nile virus increases. Loss of habitat also increases the chances for disease transmission because animals infected with potentially virulent strains are more likely to come into contact with humans as the animals' habitats are destroyed, fragmented, or bisected by roads.

WATER IS LIFE

The relationship between biodiversity and human diseases is still a controversial field of inquiry, but an undisputed scientific statement is that water is essential to life. So it's not surprising that many of the most natural natural security systems are based on water—how to drink it safely, how to store it, where it moves, how it moves, and how we move upon it. The UN estimates that one billion people lack sufficient access to potable water,[10] and a number of high-tech and low-tech technologies have been deployed to parts of the world most severely affected. But these technologies are often not accepted or don't survive the long term in the communities they were meant to serve. It turns out that prickly pear cactus, which thrive in many of the most water-starved places and are already harvested for their flesh and fruit, make excellent water filters. When ground up, their extracts can effectively filter both sediment and bacteria in water.[11]

Even when we have access to massive storage and purification systems, water is not always available when and where we need it. In the United States, for example, while the western deserts suffer through years of continuous drought, the Midwest might be flooded from seemingly ceaseless rain. Living at the edge of the Sonoran Desert in Tucson, Arizona, I've become acutely aware of this distributional problem of water. We spend most of the year panting in almost complete dryness. Then, with great fanfare of crashing thunderheads and people literally dancing in the streets with joy, the summer monsoons come like a biblical pronouncement; the sky splits open, the people retreat to their houses, and the streets turn into rivers, the parking lots into lakes. Rain cisterns made of old plastic eighty-gallon pickle barrels and large galvanized culverts turned on end, which sat empty for half a year, are overflowing within minutes. There's no way to collect it all.

Instead, rainwater "harvesting" experts talk about slowing it down by converting hard landscapes that slope toward the streets into earthen mazes that funnel water to thirsty desert plants that have adapted well to take full advantage of brief periods of heavy water and long periods of drought. This concept of slowing down and redirecting nature's excesses is at the heart of many of the ways that nature provides security to us and the ways that we squander this security.

One of the largest cities in the world, Los Angeles, California, is a madhouse of natural threats to security, according to Mike Davis, author of *Ecology of Fear,* a natural and social history of the city and its disasters that has given many an Angelino sleepless nights wondering when the next "big one" will hit (and what "one" it will be). Los Angeles's native ecology is home to earthquakes, fires, floods, landslides, and even tornadoes.[12] It has little water, and what water it gets typically comes rushing down the steep surrounding mountains after intense storms, and floods out across the entire flat Los Angeles basin. But human ingenuity quickly took care of all that. Well, it took care of the flooding problem—the earthquakes, the fires, the landslides, and the tornadoes are still there.

In fact, both the early agricultural economic base and the whole political power structure of Los Angeles was born over the control of water. Farms on the border of the original pueblo of Los Angeles were fed by *zanjas,* earthen canals, off the Los Angeles River. By the mid-nineteenth century, control of the *zanjas* rested in the hands of the *zanjero,* an appointed position whose power is belied by the fact that his salary was a third more than the mayor's.[13] If you wanted water, you talked to the *zanjero.*

Without controlling the episodic flooding of the river through canals, Los Angeles would never have grown in population as it did. But as it grew, agriculture was pushed away into the valleys of the north and east, and the *zanjas* became less important. Now the goal was to keep the flooding from taking out the factories, train tracks, and houses that were packed right up to the banks of the river. Earthen canals and the natural meandering river course were realigned into huge straight box channels lined on three sides with concrete. To make the walls high enough to contain the high waters after the rains, in some places these channels are over thirty feet tall. In one place you can stand at the bottom of a box channel and look way up to find old hobo graffiti from the late nineteenth and early twentieth centuries. It hasn't been covered up by the more contemporary gangster graffiti because it's too high, scratched out at the former base of a low overpass that once crossed the true natural bed of the river. Nowhere is the transformation of the river more stark than at the confluence of the Los Angeles River and the Arroyo Seco. The Portola Expedition of 1769, which had plied the pristine coastal lands of California, delighted when they saw these rivers come together, noting, "The beds of both are very well lined with large trees, sycamores, willows, cottonwoods, and very large live oaks," and considered it the best spot of all they had seen for "a very large plenteous mission."[14] Now this same spot is a gargantuan monument of concrete walls and freeway overpasses with nary a living thing to be seen, immortalized in the irony-laden website Friends of Vast Industrial Concrete Kafkaesque Structures.[15]

Of course, not just the river was transformed into concrete, the entire vast growing cityscape was laid out in hard surfaces—

sidewalks, parking lots, and eight-lane freeways—and former streams were redirected into concrete pipes and buried underground. Besides being hideously ugly, the net effect of all this poured concrete and asphalt (known to water engineers as "impervious surfaces") is that rather than slowing the water and letting it hydrate plants and percolate into the ground where it can later be pumped up to use, it now rushes (along with a truly astounding amount of trash, pesticides, herbicides, pharmaceuticals, and fecal coliform bacteria) straight into the Pacific Ocean. Thus, a city chronically starved of water has created the greatest system for getting rid of naturally supplied water as fast as possible, before a drop can be used.

But the slow pace of natural water management often jibes with the speed expected out of commerce and economic growth. Nowhere was this desire to optimize commerce, nor the consequences of ignoring natural security systems, better demonstrated than in New Orleans. There, the Army Corps of Engineers (the same outfit that cemented the Los Angeles River) were brought in to create the Mississippi River Gulf Outlet (MRGO), affectionately known by its acronym as "Mister GO." The lazy, meandering course of waterways from the Mississippi River delta to the open waters of the Gulf of Mexico were apparently far too burdensome for commerce. What's more, an enormous wetland system was smack in the middle of the most direct route from the river to the gulf. Already, much of the natural wetland systems surrounding New Orleans had been filled in, built upon by engineers who replaced the wetlands' storm-buffering abilities with levee walls. The Army Corps, using its most linear, most hyper-efficient mindset, simply drew a straight line across the remaining wetlands and started mixing the concrete. When the dredges finished cutting Mister Go in 1963, New Orleans had a fabulous new aquatic highway, perfect for the fast and direct passage of freight barges, oil tankers, fishing boats, and recreational watercraft, as well as salt water from the oceans (which killed the most-stabilizing wetland plants), and . . . hurricanes.

As Hurricane Katrina bore down on the southeastern United States, its strength varied between a tropical storm and a full-force Category 5 hurricane. Within the last forty-eight hours before it struck, forecasts became very clear that it would hit New Orleans. As if in preparation, Mister GO rolled out the red carpet for Katrina. With no complex topography or wetland vegetation to slow her winds or the waves of tidal and storm surge she brought with her, Katrina—although only a Category 3 storm at landfall—was able to march right into New Orleans, nearly as grand and powerful as ever.

Of course, New Orleans has always been concerned about getting flooded, being as it was constructed below sea level at the confluence of the Gulf of Mexico and the Mississippi River, the biggest drainage system in the United States. Accordingly, it has become a city behind walls, protected by a series of drainage pumps and over 350 miles of levee walls to hold back the floodwaters.[16] One of the problems with walls as a means of slowing water from its natural course is that they're not very adaptable. Floodwalls can be made to be resistant to the "100-year flood," a figure based on past performance of nature. But nature, especially now in the era of rapid human-caused climate change, isn't contingent on history. Climate, and the tropical storms fueled by a warming climate, is changing in unprecedented ways. A 100-year flood threshold calculated based on the intensity of storms in the past two decades would look much different than the same threshold calculated when New Orleans's levees were built, but the levees can't be changed so easily. Katrina wasn't the biggest hurricane to hit New Orleans in the previous 100 years, but its storm surges, up to 10 meters (33 feet), were definitely the largest.[17] The current system was designed to withstand a Mississippi River flood the size of the Flood of 1927 and a hurricane with wind conditions similar to a very strong Category 2 hurricane.[18] Sure enough, although the levees had protected New Orleans for hundreds of years, they didn't do the job during Hurricane Katrina.

The other problem with the particular levees in New Orleans is that they essentially eliminated the natural process of sediment accumulation in the Mississippi delta that provided the base layer for wetlands. This natural process had resulted in wetlands growing larger in the delta for thousands of years, a trend that was radically reversed in the twentieth century due to human activities.[19] So, on the one hand you had walls, which are not adaptable to the changes in nature, and that hand was taking away from the other hand, which was holding the best natural buffer against nature's extreme forces: the coastal wetlands.

There is a growing body of evidence, arising after disaster strikes, supporting the claim that natural, intact ecosystems are effective buffers against storm surges, flooding, and tsunamis. Typically, this is hard to study, precisely because of the variability and unpredictability of these events. Where and when to deploy experimental sensors is a tricky question, and once such sensors are deployed, scientists find themselves sitting on the ethically flimsy fence of wishing for an intense natural disaster to get some data so that maybe we'll learn more about how to protect people from the *next* disaster.

So it was something of a silver lining to otherwise grim events that Elise Granek and Ben Ruttenberg, two friends of mine conducting field ecology research in Belize unrelated to storm protection, lost a whole bunch of their experimental equipment to Tropical Storms Wilma and Gemma. Field ecologists studying interactions between organisms typically deploy a range of homemade apparatus to study different variables of interest. These might be PVC and chicken-wire cages laid over plants to keep herbivores away, or barriers bolted to rocks on the shoreline to see how well mussel beds grow in the absence of voracious starfish predators. Ecologists will typically set up several replicate plots of these cages and barriers and put them in different contrasting environments as well. Especially in coastal areas, they have to be designed to be "bomb proof"—anchored by thick galvanized hardware bolted directly into rocks—just to deal with the continual force of coastal waves.

Elise and Ben had set up a large number of these treatments in sites with natural intact mangroves and in places where coastal mangroves had been cleared. They had wanted to see how the ecology of the mangrove and the cleared coastline differed. Their experiments were going well until the storms hit and destroyed a lot of their equipment. They wouldn't end up getting much useful ecological data, but it wasn't a complete loss. Specifically, they were able to quantify which equipment was lost, and they found a difference between how many of their solidly bolted pieces of equipment were ripped away by the storms, with significantly more equipment surviving in the intact mangrove shore versus on the transformed shoreline where the mangroves had been removed.[20] It was an accidental experiment, but one that demonstrated just how much intact ecosystems buffer the forces of natural disasters.

Unfortunately, most of our tests of the protective capacity of mangroves and other natural systems destroy a lot more than experimental equipment. The 2004 Boxing Day Tsunami, which killed 600,000 people, provided a sobering overview of the power of intact coastal forests to protect from storm surges. As researchers and aid workers examined satellite imagery and the on-the-ground destruction of the tsunami, they saw a consistent pattern. Those villages behind intact mangrove forests and tree plantations survived with much less damage than those that had no coastal buffer or had converted their mangroves into large flat pools for raising aquacultured shrimp. While the largest and most intense parts of the tsunami could easily overcome even dense coastal mangroves, slightly less forceful tsunamis, which could still destroy an unprotected village, can be reduced by as much as 90 percent by coastal forests.[21]

The value of these protective services is enormous both in lives and economic infrastructure. A team of economists and ecologists estimated the value of wetlands for storm protection in the United States alone at over $23 billion per year. Yet, even as the evidence for the protective value of wetlands and mangroves accumulated, it took some time for anyone with any power to listen. Oddly enough, the 2006 Townsend "After Action" about "lessons learned" from the federal response to Hurricane Katrina doesn't contain the word *wetlands* and doesn't discuss the benefits of natural storm buffers,[22] even though much of these lands were under federal jurisdiction or had been altered by federal agencies like the Army Corps. A report by several colleagues of mine on the value of wetlands for flood control, although contracted by the government, was suppressed for well over a year. But as evidence accumulated from around the world, the voices speaking for wetlands protection were eventually heard, and they even came to dominate the discussion of post-Katrina New Orleans. Perhaps the biggest turnaround is symbolized by the Army Corps response to a congressional call for an end to "deep draft" navigation in the MRGO. In 2006 the Army Corps responded more strongly by recommending closing the MRGO to *all* traffic and restoring the surrounding wetlands, a remarkable change for an agency that for most of its existence thrived on replacing natural protective barriers with concrete.

Unfortunately, after disaster strikes we often retreat to the apparent security of technological security. After Hurricane Ike, a massively destructive storm that hit the Texas coast in September 2008, one of the first responses was to call for an "Ike Dike," essentially a huge 60-mile-long concrete wall to protect against annual hurricane season storm surges. The lack of an adaptable mindset marked both the attitudes of the Ike Dike planners and the news coverage of the dike. A former Texas mayor on the commission to study the dike's feasibility proudly proclaimed, "The elegance and the appeal of something like the Ike Dike is, with one swath, all the problems are solved." The *Wall Street Journal* article in which the mayor was quoted noted that Houston area leaders were "hoping to end the annual storm threat once and for all."[23]

THE PROBLEM WITH WALLS

The idea that walls will save us from any dynamic and changing security threat, be it storms or humans or viruses, has been proven wrong time and again. The border wall between the United States and Mexico, which cost between $1 million and $10 million per mile, slows down illegal immigrants by an estimated twenty minutes, even in its most fortified areas.[24] As former Arizona governor Janet Napolitano said in criticism of the border wall, "You show me a 50-foot wall and I'll show you a 51-foot ladder at the border. That's the way the border works."[25] What the wall has done is provide limitless cover for anti-immigration advocates who have used it as a type of political blackmail, saying they won't support immigration reform policies until the border is "secure."

From the physical world to the digital world to the biological world, hardened security systems like walls often lead to more damage inside their defenses than outside. Much of the damage in New Orleans occurred *inside* the levee walls, once the water got through. Likewise, for forty years cybersecurity experts have attempted to make "perfect" systems, but even as they continue to fail this probably impossible task (there are estimated to be 1 *million* bugs in Microsoft's Office Suite[26]) they put up firewalls as a signal that the system is perfectly secure.[27] The wall illusion quickly crumbles in the face of determined hackers who can cause considerably more damage once safely ensconced inside a walled cyber city than if there was no firewall at all.

Organisms in nature aren't immune to such breaches of their defenses. A certain jumping spider turns out to be a deadly hacker of ant "firewalls." This spider breaks into the ant nest by mimicking the unique olfactory signal of the ant colony—essentially stealing their password. Then, in a second act of deception, it gently taps ants carrying larvae, just as a fellow ant would do with its antennae. This causes the ant to drop the larvae, as if passing off the burden to another worker ant, and the spider has earned a meal.[28]

The spider's "hack" of the ant colony is mirrored in a simple and effective cyberattack that has been successful in both deliberate simulations and actual attacks. This hack involves physically scattering virus-infected USB drives in a parking lot and letting employees with security clearance inadvertently introduce the virus behind the firewall when they insert the drives into their workstations.[29]

We've mostly been talking about the two poles of nature-based natural security—either leaving nature to do its thing or making completely artificial barriers. What if, instead of making walls and diorama-like security systems that physically cannot adapt to changing conditions, we made truly living security systems? We can start by using nature itself. The storm-buffering benefits of the natural curvature of rivers can be restored. The wave- and wind-dampening effects of natural wetlands can be put back into place by creating "living shorelines" rather than hard sea walls (which actually increase coastal erosion). Sometimes the barriers to implementing these improved systems are just bureaucratic.

For example, in North Carolina, a state blessed with abundant wetlands but cursed with being incredibly vulnerable to sea-level rise, living shorelines in front of wetland properties are not as popular as putting up hard cement walls (which make up 79 percent of the constructed barriers to flooding) mostly because building a concrete sea wall requires only a relatively simple "general permit" from the state, whereas constructing a living shoreline requires an odious and costly "major permit."[30]

The benefits of these living systems are not trivial but accrue at many different levels, a hallmark of natural security systems. Consider that production of cement used to build sea walls creates a huge amount of greenhouse gases, while wetlands serve as a sink for greenhouse gases. Wetlands are also effective water filters. My own research on a tiny remnant wetland being restored in Los Angeles shows that even small wetlands help remove human pathogens,[31] and wetlands also serve as a nursery for important fish species that attract a huge diversity of bird species. In turn, as I discussed earlier, a high diversity of bird species has recently been shown to be a key factor in reducing transmission of some diseases such as avian flu. That means we can create security systems that protect us from intensifying storms but also reduce our climate impact, protect our water supplies, increase biodiversity, improve coastal fisheries, and lessen the impact of infectious diseases. Can a cement wall do all that?

Indeed, essentially the whole living Earth is built on these synergistic cooperative arrangements and through them becomes a self-regulating system. This system isn't static—I've argued throughout this book that biological systems are changing at every level all the time—but it is constrained within boundaries that have supported some kind of life for billions of years and life that is fairly like our own for at least several hundred million years.

This kind of dynamic stability requires balance—a temperature neither too hot nor too cold, abundant oxygen for metabolism but not so much that everything instantly catches on fire, enough available nutrients for primary productivity (the algae and other photosynthetic organisms at the base of most food webs) but not so much to overpopulate an area to the point of eutrophication (the drawdown of oxygen due to the decomposition of too many primary producers in a given area). This is the basis of James Lovelock's "Gaia hypotheses"—that the Earth as a whole is a self-regulating system that supports life and evolution. Some have overextended the Gaia idea—giving the Earth mystical powers to control Her own development, or suggesting that the purpose of green plants is to support the herbivores that eat them or the animals that breathe the oxygen they produce—but there is evidence to support the basic idea that interactions at small scales can build into an essentially self-regulating complex system.[32]

Although these complex self-regulating systems don't have any goals, they do tend to protect components of the system, especially those, like humans, that are adapted to a range of conditions. This is the element of "resilience" or "robustness" that people who study complex systems are always looking for. Resilient natural systems protect us—not just from instant planetary conflagration through unchecked oxygen production but from a range of other security threats; from individual infections to epidemics and from a house flooding to a city drowning.

These are the "nature's services"[33] or "ecosystem services" that refer to the many things, essential to society, that nature provides to us. These include the pollination services of insects, birds, and bats, as well as the nursery habitat for economically valuable fish stocks provided by wetlands and mangroves.[34] The list goes on: blockage of damaging UV radiation by ozone in the upper atmosphere.

Conversion of dead organisms into organic fertilizer. Chemical-free control of pests like mosquitoes. Key ingredients for medicinal products. Recreational opportunities afforded by intact ecosystems. The aesthetic value of a pristine wilderness.

So, how much are all these services worth to us? A lot. There has been some controversy over exactly what "a lot" means. You can imagine how difficult it would be to put a price on some things—for example, add up the total sold value of all the paintings, novels, photographs, and poems inspired by nature, then develop and add in some kind of "happiness" quantity related to human appreciation of nature (to capture the value to the many nonartists and nonwriters, and to the many artists and writers who never sell a thing) and for good measure add in the reduced cost of medical care for the reduced stress and improved breathing that comes through interactions with intact nature. That crazy exercise (which economists try to do, I guess because economists never really have to get anywhere close to a real number for things) might give us a ballpark estimate of just the aesthetic value of ecosystem services. Probably the most accurate statement of the value of ecosystem services is that we don't know how much they are worth, but we know they are worth a lot more than nothing, which is how we currently value most of them.

The same could be said of "natural security" in general. With countless individual organisms and billions of years of evolutionary history to look at, we have only made the very beginnings of a much deeper journey back into the natural world. But just as not having a precise value for the Earth's ecosystem services is no reason to ignore them altogether, not having all the answers from nature at our disposal is no reason not to begin transforming our society into a more adaptable organism. In the concluding chapter, I will outline the first steps. If taken correctly, they will generate their own momentum and lead us down a path toward ever greater adaptability.

CONCLUSION: THE END AND THE BEGINNING

At the beginning of this book I said that adaptation arises from leaving (or being forced from) one's comfort zone. Accordingly, it's understandable that we might be a little resistant to dive into this strange world where reacting to the previous crisis is no longer good enough and making vague predictions of the future no longer counts as "doing something." It's natural that we'd come up with all sorts of excuses for why we can't be more adaptable. But one of the results of using nature—with its relentless ability to solve problems and neutralize unpredictable threats—as a template for adaptability is that it weakens almost every excuse we have for not becoming more adaptable. The overwhelming success of adaptation in nature practically shames us into at least trying. And everything that seems like a barrier to change has already been crossed in nature. We complain that our bureaucracies are too institutionalized to change, but even organisms whose outer appearance has remained steadfastly unchanged for millions of years can be highly adaptable by farming out that adaptable capacity to semi-independent parts, like immune cells and skin color pigment cells. We argue that there are people we just can't work with or who will never come to peace with one another, but in nature the meekest organisms form beneficial symbiotic relationships with the most terrifying. We argue that we can't have guns *and* butter, but every successful living thing already knows how to balance a way to defend itself, a way to nourish itself, and a way to reproduce itself. Perhaps the last remaining excuse is that we are just different altogether from every other living thing. At a mechanical, biochemical level, this can't be true. Under our hoods is the same DNA engine that powers most life forms. We may have neat flame decals and a clever dashboard GPS system, but we're essentially the same type of vehicle weaving through traffic (and occasionally driving right over it) on the road of life.

Nevertheless, when we take a broader *naturalist's* view of how different we appear from all other organisms, several contradictions arise. We often seem to be removed from the relentless force of natural selection. After all, an overweight nearsighted weakling can still successfully procreate. But at the same time, we can't deny that right now billions of people in fact do face a daily struggle to survive. We also seem to be controllers of our own evolution. We have essentially turned ourselves into the highest flying, deepest diving, fastest moving animals on Earth. Yet, the forces of nature—be they shrieking hurricanes or microscopic microbes—can still exert enormous influence on where and how we live.

The contradictory nature of these observations reveals both why an evolutionary approach to security is so important and why it is more attainable than we might think. We have in effect insulated ourselves from the forces of natural selection that in nature provide for the longevity of adaptable organisms, and this may have ironically weakened us in some ways. Yet, we are fortunate in that given the right information we can actively, deliberately, and quickly adapt our responses to threats in our environment, a luxury that few other organisms possess.

But this, then, brings us back to the ultimate contradiction. That is, how can we be so wonderfully adaptable as individuals—whether living among the beasts in ancient African savannahs or turning the tables on airline hijackers who made it through dozens of layers of homeland security—and yet create organizations and institutions that seem so nonadaptable? Moreover, on a daily basis we abide by (voluntarily or through various forms of coercion) these institutions—obediently removing our shoes for airport security, loyally paying taxes to a system that seems vastly overinvested in far-off security threats rather than the ones right here in our communities, and dutifully showing up to work for agencies and corporations that we know are highly resistant to change. We are well aware that in emergencies, and in intense survival situations like warfare and in trauma centers, our adaptive capacities come out and shine—they break through the institutional straightjackets we've allowed ourselves to wear. What we seem much less sure of is how to actually harness the power of adaptability and use it at will. How can we, *unlike nature,* deliberately design systems and ways of dealing with security problems that, *like nature,* have the ongoing organic capacity to automatically respond to and overcome problems as they arise?

The key is creating an *adaptable cascade,* in which some small adaptable actions set off other adaptable actions that ultimately lead to a system that generates its own momentum toward ever more adaptability. You won't find a nicely branded "adaptable cascade" in nature—where one adaptation naturally cascades into another. But in human society, where we've continually put up barriers to adaptation, we need a process to help our adaptive capacities flow again. An adaptable cascade brings with it all of the components of natural adaptive systems that I've discussed in this book. It creates a decentralized organization of multiple semi-independent problem solvers. It accelerates learning by selecting for success. It creates useful redundancy, which it then uses to take control of uncertainty. It helps facilitate symbiotic partnerships. And it provides a better pathway for protecting the natural ecosystem services that protect us.

It's lovely if you can create an innovative organism powered by adaptive cascades as a complete unit. The founders of Google did their best to design adaptability into their business model, and even as a multi-billion-dollar company that serves billions, they've managed to keep the company infused in adaptability. The vast majority of us don't have the luxury of creating a whole new organism, though. Fortunately, the deluge of an adaptable cascade can be unleashed safely in any business, any bureaucracy, any police precinct or fire department, any household. Moreover, we are not jumping off this cascade into the void inside a barrel. We have incredible support systems at our disposal. First, nature itself is a powerful security system that works even better when paired with adaptable approaches to human problems. Second, keep in mind that we are products of evolution and adaptation ourselves. Our natural state is one of adaptability. Sometimes we need to let nature take its course, and we'll be much more able to do this if we ourselves are heading down an adaptable path.

The first step for creating an adaptable cascade isn't to create a new report on how to make the organization more adaptable or to hang "Be Adaptable!" banners or to give out "Just Do Good Enough!" lapel pins. It is to transform whatever sounds like an order now into a challenge and to create whole new challenges. An order is anything created by a small elite group (or powerful individual) that is forced upon anyone else in the group under the expectation that it will be followed to the letter. A challenge, by contrast, is an open solicitation for help to solve an identified problem. Issuing a challenge is not about relinquishing control or completely overturning an existing hierarchy. The person or group issuing the challenge still has the power to design it, shape the incentives that will attract people to it, set the rules, and determine who will get to participate in the challenge. In other words, switching to a mode of issuing challenges doesn't have to radically alter the structure or power dynamics of a given organization. At the same time, it can radically alter how problems are solved.

Challenges work because they emulate the natural adaptive organization of nature, where multiple semi-independent agents are solving problems where they occur. In more human terms, they give ownership of a problem to the people who have to work on solving it. Encouraging a higher degree of autonomy in problem solving is particularly important in human organizations where power dynamics may constrain the space for problem solving—the well-known problem of having too many "Yes Men" who feel compelled to agree with the boss.

As a challenge is initiated and multiple independent problem solvers take part in it, the need to issue advice and orders begins to fall away naturally. For example, I could directly advise in this book something pithy like "don't build wall-like security systems," but I strongly suspect that agencies and organizations in the midst of an adaptive cascade will quickly dispense with static security systems as multiple problem solvers look at them and say, "Wait a second, I could figure out an easy way to get around that."

Indeed, as individuals work on the challenge and share their results, learning outcomes as a whole will improve as an adaptive cascade takes its course. Like learning in nature, learning will happen automatically as an organization becomes more adaptable. Too many business books and corporate seminars and consultants try to institutionalize learning, but if your organization isn't learning, it's not because it hasn't discovered some playbook on how to learn. Any plan for learning will become redundant as an organization becomes adaptable.

At a university, that learning plan is called a syllabus, and it's what every instructor is expected to hand out to students on the first day of class. It wasn't until I thought about setting off an adaptable cascade in my own classroom that I recognized how much the syllabus, the stalwart learning plan edified into university life, was constraining learning. For years, as a university instructor, I became increasingly frustrated about how little effort students put into participating in classes I taught. Even when I did things like have a student lead class sessions, I found that all the rest of the students tuned out to the student as much as they did to me. So, after thinking about how the lessons of adaptability might apply to teaching, I started running my classes based on challenges.

The first thing I did when I entered class for the first time in the semester was have the students rip up the syllabus I was required to give them. Then I asked them to respond on 4x6 index cards to two challenge questions: (1) "What do you want to learn in this class?" and (2) "What can you share from your experiences that will help us learn about these topics?" By combining their responses to the two challenges, we were able to create a whole new syllabus, but one that everyone had participated in, rather than one handed down from a single professor. The challenge wasn't a complete free-for-all: all of their responses needed to have something to do with the course topic, marine conservation. And the challenge didn't stop with making the syllabus. By midnight on the night before every class period each student had to contribute something related to the class topic to an online Wikipedia-style website. When class came, I found I never had to pull teeth to get the students to talk; because they owned not only the course but also a little piece of every class meeting, they were eager to share their contributions and to draw out connections.

And another surprising thing happened. I learned far more during the semester than I had in preparing and running any previous course. Had I not challenged all the students to make the class their own, I would never have seen the firsthand view of one student who grew up in the canal zone in Panama and shared with us how international trade, global politics, and massive-scale ecological transformation interacted there. I would have never really understood the difficulty of trying to run a closed-cycle sustainable fish farm as one student's father struggled to do in her home country of Venezuela. I would never have learned, as I did from a very environmentally conscious Chinese student, the difficulty of staging a wedding there that did not include shark fin soup on the menu, as his own wedding had scandalously omitted.

The only real requirement for learning that may need to be deliberately institutionalized is a commitment to learn from success. Many business consultants continue to peddle the cheap wisdom of "learning from failure" as some sort of innovation, but most learning from failure will become merely, and at best, a single solution to a problem that's already occurred. In nature, innovation comes first, and learning accrues from successful innovations, which in turn allows an organism to survive and continue to innovate. For us to replicate this natural process, we must amplify, reward, and replicate successes small and large, as they occur.

But even this artifice can be kept fairly minimalist. Example after example shows that if an adaptable problem-solving system is in place, the incentives to bring out and replicate success are almost laughably immaterial. Employees who help green 3M as part of its Pollution Prevention Pays program, which has reduced over a billion tons of pollution since its inception in the late 1970s and has saved the company over $3 billion, get little more than a statement of recognition from corporate headquarters. Yet, to date over 8,100 employee-directed pollution reduction projects have been recognized by 3M.[1] The Stanford engineers who created Stanley received bragging rights and a cash prize that was likely less money than they spent on the project. When people are given the power to create a more adaptable system, their participation and sense of ownership in that system that is working well is often all the motivation they need to keep on innovating.

When learning is taking its natural course among multiple semi-independent problem solvers, multiple opportunities for creative redundancy are created. Stanford's Stanley won DARPA's Grand Challenge, but if Stanley didn't run that day, there were a number of other entries that each solved the challenge in a different way and were ready to take the prize. In the early 1700s, the English Parliament set up a challenge to solve the pernicious problem of identifying the correct longitude while at sea. While the conventional wisdom said that astronomers would develop the solution based on celestial navigation, it was fortunate that Parliament didn't just assemble a group of the world's best astronomers to try to solve the problem behind closed doors. In the end, while astronomers and sailors made some important contributions, the winning solution came from a watchmaker who designed a time piece that could keep extremely accurate time even during the rigors of an ocean voyage.[2]

Embracing creative redundancy increases diversity, which is one way to get control of uncertainty, and even to turn uncertainty into an advantage. Consider what Phil Jackson, one of the most successful basketball coaches of all time, did with Dennis Rodman, one of the most unpredictable basketball players of all time. Rodman might show up to a game with his hair dyed bright green and his nails painted to match, and to a press conference in an evening gown. While most coaches and sportswriters considered Rodman a massive liability, a distraction that threatened to destabilize the tightly wound system of a successful team, Jackson embraced Rodman's personality and made it part of a Chicago Bulls franchise that won three straight championships. He publicly affirmed the value of Rodman's idiosyncrasies, bestowing upon him the role of a Heyoka, a Lakota Indian trickster spirit who cross-dressed and did things backwards to expand the consciousness of his fellow people and help them understand their own place in the world.[3] Lost to many observers in the distraction of Rodman's hair color and odd lifestyle was the fact that he was an unmatchable defensive force who took the little-heralded job of rebounding loose basketballs to a whole new level. Ultimately, although Jackson coached some of the game's most predictable superstars—Michael Jordan, Shaquille O'Neal, and Kobe Bryant—he still considers Rodman the greatest player he ever coached.[4]

In this light, one of the biggest concerns about issuing open challenges—that there is a lot of uncertainty about what will come back in response—can be seen as an asset. The uncertainty that multiple problem solvers bring with them is its own form of naturally emerging diversity, which provides rich ground for adaptation. Although Rodman was impossible to hide in the Bull's lineup, there very well may be a Heyoka hiding in your world that hasn't been given the right challenge yet.

Invariably, new symbiotic partnerships will be borne out of the adaptive cascade. Symbiotic partnerships are essential to adaptation, and they extend any adaptive capacity. Yet creating these partnerships is also something that is not done well as a mandate from the top down. The government has mandated many "interagency task forces" and the like that were designed to create partnerships between agencies that rarely talked to one another. But with a narrow set of allowable tasks and a required number of annual meetings, these task forces tend to become exercises in which representatives from each member agency who have little power to make decisions come to check off a box. Real symbiosis arises automatically when different entities find out that they can solve imminent problems better together than they could on their own. As in nature, these symbioses can grow out of competition. For example, University of Washington professor David Baker created an online game to help solve a longstanding challenge to understand the structure of a protein related to HIV. After the winning team solved this 10-year-old mystery in just 10 days, Baker noted, "Competitive social interaction is a very strong driving force."[5]

Symbiotic relationships may end as quickly as they were formed, or they may become long-term partnerships, if both parties find at least some additional benefit from staying together. Accordingly, Baker's team has now issued additional challenges to the online gaming community, just as DARPA maintained communities of problem solvers by issuing ever more complex challenges. In this sense the central challenge of the adaptive cascade becomes the catalyst that new symbiotic partnerships are built around. The more perspectives that are brought in to address the challenge, the more opportunities to develop new symbiotic relationships emerge.

Finally, even the natural security services provided by nature are likely to be better protected under an adaptable cascade. Environmental protection in the future is unlikely to come about from yet another fundraising campaign from World Wildlife Fund or through sales of Sierra Club calendars. As Paul Hawken documented, small localized groups around the world have been far more effective at protecting the ecosystem services in their regions than the huge multinational environmental groups. This is great for protecting a local mangrove forest, but can this localized power deal with global climate change? It can if it's linked to larger-scale information and resources, and this is the nature of adaptable cascades—they merge the power of localized problem solvers with a centralized organization that is better able to see the big picture and has the resources needed to solve problems.

An adaptable approach to environmental protection that inherently makes this connection is found in the concept of the "public trust doctrine."[6] The doctrine, which is fundamental to our formation as a country, and may date back to the Magna Carta,[7] states that natural resources (like the wetlands that should be protecting New Orleans) belong to all citizens equally and are only held in trust by the government, which has an obligation to protect, grow, and repair the body of the trust. In contrast to the current system of lobbying groups working to create (or fight against) fairly static conservation laws that become solely the province of government to enforce or not, public trust governance creates the opportunity for adaptable, feedback-oriented conservation.

The challenge put forth by public trust governance is, How do we balance extractive uses of the trust that provide immediate benefits, with the obligation to protect the trust for the future? The solution to the challenge lies in both a central government that can keep track of the whole portfolio of trust resources, and a citizenry that is able to, passively or actively, contribute to trust management through each individual's valuation of trust resources. Peter Barnes, for example, has argued that the atmosphere (and the service of climate stabilization it provides) is a public trust and, as such, all citizens should be compensated for activities that damage the trust through a fund that polluting industries must pay into.[8] This is not unlike the state of Alaska, which provides annual royalty checks to all Alaskans from a fund paid for by oil and gas developers. The concept of the public trust is the catalyst for a symbiotic partnership between multiple individuals with an incentive to solve a challenge and a central organization that has the resources and power to put the solution into practice.

A nature-mimicking adaptive cascade is modular, just like nature. Thus, we can initiate it in whatever module of society we happen to work within. Likewise, because nature is hierarchical, adaptive cascades can be created to fill various needs across scales of time and space. The initial challenges, for example, can be part of a five-year reorganization of the whole organization, they can be aimed at the next quarterly reporting period, or they can be renewed daily. When I visited Google's campus in Mountain View, California, I was amused to find nearly incomprehensible (to me) daily programming and engineering challenges posted on sheets in the bathroom stalls. If the incentives for learning from success are properly set up, adaptable systems will spread throughout an organization, regardless of where they were originally initiated.

ADAPTABILITY WAY BEYOND SECURITY

I am hoping that by this point in the book you are thinking not just about what you've read, but about what you haven't read. Security is both too broad a topic to cover in one book on adaptation, and too narrow. There are myriad security-related problems I haven't touched on here. Examples that might appear as glaring omissions (depending on your own background and political positions) include the sorry state of our critical infrastructure, the very real specter of a debilitating cyberattack, or growing income and educational inequalities within and between nations. My goal with this book was not to take a survey of today's security problems (which may or may not be tomorrow's problems), but developing a framework for dealing with security problems, no matter what they are, where they arise, or when.

Which gets at the fact that security is also too narrow an application to absorb all the adaptable lessons of nature. I focused on security questions because when I started this project back in 2002 in Washington, D.C., the concept of security was so strongly manifest that it was barely possible to think of anything else. I don't still keep a chemical evacuation mask under my desk, but there remain abundant reminders that we live in an unsecure world. As I write this, the Mississippi River is coming to its highest rise since the Great Flood and the Army Corps of Engineers are quickly breaching the levees and spillways protecting thousands of homes and farms to avoid a down-river breach of the levees protecting millions of people in Baton Rouge and New Orleans and the oil refineries and chemical plants beyond.[9] Osama bin Laden has been killed, but days later al-Qaeda killed sixty people in a dual-suicide reprisal attack.

At the same time, most Americans are currently thinking more about job security than terrorism. Scientists are fretting more than ever over the unchecked consequences of climate change. And the daily things that we deal with—getting our kids to school in the morning, working with aggravating colleagues, trying to balance income and expenses, seem far removed from security policy writ large.

The biological concepts of adaptability that I've been looking at in the context of security are just as applicable to the non-security-related parts of our brains and our lives and our institutions. The same fundamental linkage between biology and security that I established at the outset of this book—that biological organisms and human societies both face highly variable and highly unpredictable threats—applies to almost everything we are involved in as humans. For example, your three-year-old's moods, the value of your stock portfolio in three years, and the ecological condition of the herring stream you've been fishing since you were a kid are all things that will change radically and unpredictably in the near future. Likewise, at a different level of human organization, school boards want to know if their populations of children will perform well enough on tests to earn more federal funding, brokerage firms want to know if their investments will put them in the black, and fisheries management agencies want to know how herring populations overall will do over the next few years. We would love to be able to predict these changes, but we can't. Like the stock market, we also can't rely on past performance as a guide to the future. We've got to instead find that sweet spot where we can not only survive, but thrive, in a changing world where tomorrow's problem is almost certain to be different than yesterday's. We get to that sweet spot by developing adaptability in ourselves and in our institutions.

Bringing the full scope of biological adaptation to our own lives takes us to a crossroads. We can become overwhelmed by the enormous tangle of life around us and retreat back to more familiar problem-solving pathways, or we can continue to plow forward into the complexity and try to make some practical sense of it, even if we don't understand the whole. As a biologist, I've already chosen my path—I know that I will never understand biological systems the way an accountant can master the tax code. But I also know that even a partial understanding of biology yields surprising insights, especially in an era when we've largely forgotten about nature's power to expand our minds. We may no longer have the luxury to cruise the world's oceans for five years like Darwin, or revel in our first summer in the Sierra like John Muir, or spend a brooding solitary winter in an outermost cabin like Henry

Beston. But nature is still there. Its diversity is still staggering, its mysteries still profound. And its lessons are still free for the taking, completely open source and unclassified. It's time to feel the cactus spine, listen to the marmot's shrill call, and stare deep into the eye of an octopus.

ACKNOWLEDGMENTS

I AM GRATEFUL TO the John Simon Guggenheim Memorial Foundation for a fellowship that was critical in helping me complete this book. I thank Diana Liverman, Jonathan Overpeck, and the staff of the Institute of the Environment at the University of Arizona for their support of my work.

Writing a book is built on relationships, and it builds relationships as well. I only regret that here I can only thank a subset of the many relations that helped build and were built from the creation of this book.

I have been incredibly fortunate to have three academic mentors who have remained sources of inspiration and constructive criticism throughout my career. Don Kennedy, who was among the first people I shared this idea with, has been a cheerful enabler of some of my most outlandish ideas. From helping me investigate the "EZ Cheez" loophole in ozone-depleting chemical regulations during an undergraduate semester he taught in Washington, D.C., to the current work, he has always helped me focus the right questions and talk to the right people. Steven Gaines, dean of the Bren School of Environmental Management and my graduate advisor, taught me through example the craft of turning a tricky problem on its head to look for an unexpected solution. I can never say enough about Chuck Baxter, my first scientific mentor, ordained Universal Life Church minister at my wedding, and supplier of insanely good Carmel Valley–grown raspberry pepper jam. Although long retired as a biology lecturer, his current explorations of the subconscious mind have been highly influential on my thinking here.

The National Center for Ecological Analysis and Synthesis, with funding from the National Science Foundation, supported the initial working groups that led to many of the ideas discussed in this book. It is fortunate that my initial proposal was reviewed by Professor Larry Crowder, a boundless mind who later became a colleague and friend, as he had the nerve to advance my proposal when everyone else thought I was out of my mind. I am grateful to the other limitless minds from many fields of science and security practice who took a risk and joined what became some of the most stimulating conversations I've been a part of in my professional career. Too many people have contributed to these and subsequent "Natural Security" working groups I've organized to list here, but a small core has continually been a part of this project from the start and continues to enthusiastically explore new research projects on adaptability and security. Geerat Vermeij and Luis Villarreal not only provided astoundingly astute contributions to the working groups but also allowed me to share their unique personal stories. Scott Atran frequently reported back from the field, and whether he had been talking to disaffected youth of the Near and Middle East or getting grilled by congresspeople in D.C., he always had something interesting to share. Marmot-loving UCLA behavioral ecology professor Daniel Blumstein and University of Edinburgh Reader in International Relations Dominic Johnson are both brilliant at drawing out the counterintuitive and hidden solutions to complex problems. Most especially, I am grateful for the continual guidance and friendship of Dr. Terence Taylor, whose background as a soldier, intelligence officer, and weapons inspector could not be more different than mine, but who immediately recognized the synergy of bringing our different approaches to "living with risk" together.

The Center for Homeland Defense and Security at the Naval Postgraduate School in Monterey, California, just down the road from the marine biology lab where I got my start as a scientist, has been an invaluable source of inspiration and feedback for the ideas in this book. Both the faculty and participants in their Masters in Homeland Security and Executive Leadership Program courses exemplify the willingness to step away from the norm that will be necessary to bring adaptable ideas into practice. In particular, I am grateful to Chris Bellavita for his great enthusiasm for unusual approaches to everything, to lecturers Sam Clovis and Frank Barrett for their unique perspectives, and to Heather Issvoran for promoting my work there. Former student and U.S. Air Force Major Noel Lipana was very insightful in sharing his experiences with counter-IED efforts in Iraq, work that undoubtedly saved many soldier and civilian lives. Along similar lines, the Office of Naval Research Global was forward-thinking in organizing a conference in June 2010 on "operational adaptation" in Edinburgh, Scotland. The civilian and military participants in that conference produced an inspired discussion and debate about adaptability that undoubtedly shaped my views here. It was there that I met Major Douglas Cullins, who is a model for how to be both a textbook "clean marine" and a highly adaptable leader.

My editor at Basic Books, T. J. Kelleher, "got it" from the start. His deft hand made revising relatively painless, even for my stubborn brain. My literary agent, Esmond Harmsworth, continually amazed me with how hard he worked on my behalf. He pored through every proposal and every page of this book with his savagely sharp analytical mind and made it better on all counts.

My father-in-law, Chester Crocker, has forged a career building peace out of some of the world's most intractable conflicts. Beyond the wisdom he shared with me directly and through selections from his seemingly infinite reading list, the respect his name carries across the political spectrum opened many doors and established many relationships vital to the completion of this book.

Finally, the deepest lesson from my scientific field of ecology, and not coincidentally the most important lesson of adaptability in this book, is that relationships matter more than anything. My closest relationships—my parents, my wife Rebecca, and my daughters—are beautiful manifestations of the creativity, commitment, and love necessary to nurture a project like this. Thank you.

Rafe Sagarin
Tucson, Arizona

NOTES

PROLOGUE

1. Mott, Maryann. "Did Animals Sense Tsunami Was Coming?" National Geographic News. January 4, 2005. *http://news.nationalgeographic.com/news/2005/01/0104_050104_tsunami_animals.html*; and "Pre-tsunami Animal Behavior." *http://www.freewebs.com/asiadisaster/unusualanimalbehaviour.htm*. Both accessed March 29, 2011.

2. MacKinnon, Ian. "Aceh Residents Disable Tsunami Warning System After False Alarm." *Guardian,* June 7, 2007. *http://www.guardian.co.uk/world/2007/jun/07/indonesia.ianmackinnon. Nizza, Mike. "To Break a Tsunami Alarm." New York Times, June 8, 2007. http://thelede.blogs.nytimes.com/2007/06/08/warning-systems-unplugged/*. Both accessed March 29, 2011.

3. Public/Private Fire Safety Council. "Home Smoke Alarms and Other Fire Detection and Alarm Equipment." White Paper. 2006.

4. "Troops Grill Rumsfeld over Iraq." *http://news.bbc.co.uk/2/hi/middle_east/4079201.stm*. "Rumsfeld Gets Earful from Troops." *http://www.washingtonpost.com/wp-dyn/articles/A46508–2004Dec8.html.*

"Soldiers Must Rely on 'Hillbilly Armor' for Protection," *http://abcnews.go.com/WNT/story?id=312959&page=2*. All accessed April 13, 2010.

5. "Rumsfeld Set Up: Reporter Planted Questions with Soldier." *http://www.freerepublic.com/focus/f-news/1297858/posts*. Accessed April 13, 2010.

6. IED casualty figures compiled from icasualties.org.

7. Personal communication from reporter Nir Rosen, January 29, 2010.

8. CRS Report. Andrew Feickert. August 3, 2009. Mine-Resistant, Ambush-Protected (MRAP) Vehicles: Background and Issues for Congress. RS22707.

9. Moll, R. J., J. J. Millspaugh, J. Beringer, J. Sartwell, and Z. He. "A New 'View' of Ecology and Conservation Through Animal-borne Video Systems." *Trends in Ecology & Evolution* 22 (2007): 660–668.

CHAPTER ONE

1. Davis, M. *Ecology of Fear: Los Angeles and the Imagination of Disaster*. New York: Vintage, 1999.

2. Food and Agriculture Organization of the United Nations. "Hunger in the Face of Crisis." In *Economic and Social Perspectives.* United Nations, 2009.

3. *http://www.infidels.org/library/historical/charles_darwin/voyage_of_beagle/*; and Darwin, Charles. *The Voyage of the Beagle: Anniversary Edition.* Washington, DC: National Geographic, 2009.

4. Stott, Rebecca. *Darwin and the Barnacle: The Story of One Tiny Creature and History's Most Spectacular Scientific Breakthrough*. New York: W. W. Norton, 2003.

5. The Penguin Books Log of the Sea of Cortez, p. 124.

6. Henderson, Donald A. "The Eradication of Smallpox." *Scientific American* 235: 25–33 (1 October 1976). doi:10.1038/scientificamerican1076-25.

7. Wulf, W. A., and A. K. Jones. 2009. "Reflections on Cybersecurity." *Science* 326: 943–944.

8. Larson, B. M. "The War of the Roses: Demilitarizing Invasion Biology." *Frontiers in Ecology and the Environment* 3 (2005): 495–500.

9. See, for example, *http://www.drugpolicy.org/library/factsheets/effectivenes/index.cfm* on the war on drugs (accessed August 30, 2010); and Crocker, Chester A. "A Dubious Template for US Foreign Policy." *Survival* 47, no. 1 (2005): 51–70, on the war on terror.

10. "Where's the Remotest Place on Earth." *http://www.newscientist.com/gallery/small-world.* Accessed April 15, 2010.

11. Levin, S. A. *Fragile Dominion.* Cambridge, MA: Perseus, 1999.

12. Binnendijk, Hans, and Richard L. Kugler. "Adapting Forces to a New Era: Ten Transforming Concepts." In *Defense Horizons.* Washington, DC: National Defense University Center for Technology and National Security Policy, 2001. National Public Radio, "Interview: U.S. Army Brigadier General Joseph Votel and U.S. soldiers describe looking for, finding, and destroying IEDs in Iraq" (National Public Radio, 2005). Personal comments to the author from TSA, Coast Guard, FEMA, and other DHS agents.

13. Ripley, Amanda. *The Unthinkable: Who Survives When Disaster Strikes—and Why*. New York: Crown, 2008.

CHAPTER TWO

1. Food and Agriculture Organization of the United Nations. "Hunger in the Face of Crisis." In *Economic and Social Perspectives.* United Nations, 2009.

2. See, for example, NOAA Technical Memorandum NMFS-SEFC-278.

3. Robaid.com. "Mussel Biomimicry Could Lead to New Super-Strong Polymers." March 5, 2010. *http://www.robaid.com/bionics/mussel-biomimicry-could-lead-to-new-super-strong-polymers.htm.* Accessed August 30, 2010.

4. Feder, M. E., and G. E. Hofmann. "Heat-Shock Proteins, Molecular Chaperones, and the Stress Response: Evolutionary and Ecological Physiology." *Annual Review of Physiology* 61 (1999): 243–282.

5. Shillinglaw, S. *"Introduction." Cannery Row.* New York. Penguin Books, 1994. Pp. vi–xxvii.

6. *http://video.google.com/videoplay?docid=-7004909622962894202#*. Accessed April 7, 2010.

7. *http://scienceblogs.com/notrocketscience/2009/12/octopus_carries_around_coconut_shells_as_suits_of_armour.php.* Accessed April 7, 2010.

8. Ricketts, Edward F., Jack Calvin, and Joel W. Hedgpeth. *Between Pacific Tides*. Stanford, CA: Stanford University Press, 1985. Pp. 175–176.

9. Ricketts, Edward F. Outer Shores Transcript. June 27, 1946. In Rodger, Katharine A., ed. *Breaking Through: Essays, Journals, and Travelogues of Edward F. Ricketts.* Berkeley: University of California Press, 2006.

10. *http://www.pbs.org/wnet/nature/interactives-extras/animal-guides/animal-guide-blue-ringed-octopus/2177/*. Accessed April 7, 2010.

11. Tamm, E. E. *Beyond the Outer Shores: The Untold Odyssey of Ed Ricketts, the Pioneering Ecologist Who Inspired John Steinbeck and Joseph Campbell.* New York: Thunder's Mouth Press, 2004.

12. Rodger, Katharine A., ed. *Breaking Through: Essays, Journals, and Travelogues of Edward F. Ricketts*. Berkeley: University of California Press, 2006.

13. Ricketts, Edward F. Outer Shores Transcript. June 26, 1946. In ibid.

14. Ricketts, Edward F. "The Philosophy of Breaking Through," 1940. In ibid.

15. Ricketts, Edward F. Zoological Preface. August 27, 1940. Accessed at Stanford University Libraries Office of Special Collections, Stanford, CA.

16. Allee, W. C. *The Social Life of Animals.* New York: W. W. Norton, 1938.

17. Wilson, E. O. *Naturalist*. New York: Warner Books, 1994.

18. Dayton, P. K., and E. Sala. "Natural History: The Sense of Wonder, Creativity, and Progress in Ecology." *Scientia Marina* 65 (2001): 199–206; and Dayton, P. K., "The Importance of the Natural Sciences to Conservation." *American Naturalist* 162 (2003): 1–13.

19. Marshall, S. "Applying Evolutionary Concepts Outside Biology." *Trends in Ecology & Evolution* 24, no. 8 (2009): 412–413.

20. Ibid.

21. Vermeij, G. *Nature: An Economic History.* Princeton, NJ: Princeton University Press, 2004.

22. May, Robert M., Simon A. Levin, and George Sugihara. "Ecology for Bankers." *Nature* 451 (2008): 893–895.

23. Nesse, R. M., and G. C. Williams. *Why We Get Sick: The New Science of Darwinian Medicine.* New York: Vintage, 1994. Armbuster, Peter. "The Sun Rises (Slowly) on Darwinian Medicine." *Trends in Ecology and Evolution.* 23, no. 8; 422–423.

24. Benyus, J. *Biomimicry: Innovation Inspired by Nature.* New York: William Morrow, 1997. See also *http://www.biomimicryinstitute.org/*.

CHAPTER THREE

1. Geddes, Lisa. "I'm Planning to Throw Rocks at You." *NewScientist.* March 14, 2009, p. 10.

2. Roach, John. "Newfound Octopus Impersonates Fish, Snakes." National Geographic News. September 21, 2001. *http://news.nationalgeographic.com/news/2001/09/0920_octopusmimic.html.* "Mimic octopus." *http://en.wikipedia.org/wiki/Mimic_Octopus.* Both accessed July 14, 2010.

3. LePage, Michael. "Gene Machine." *NewScientist.* November 22, 2008, pp. 44–47.

4. Ten Cate, C., and C. Rowe. "Biases in Signal Evolution: Learning Makes a Difference." *Trends in Ecology & Evolution* 22, no. 7 (2007): 380–387.

5. Spinney, Laura. "Tools Maketh the Monkey." *NewScientist*. October 11, 2008, pp. 42–45.

6. Kenney, Michael. *From Pablo to Osama: Trafficking and Terrorist Networks, Government Bureaucracies, and Competitive Adaptation*. University Park: Pennsylvania State University Press, 2007. Pp. 82–83.

7. Gopnik, Alison. "When We Were Butterflies." *NewScientist*. August 1, 2009; and Gopnik, Alison. *The Philosophical Baby: What Children's Minds Tell Us About Truth, Love, and the Meaning of Life*. New York: Farrar, Straus, and Giroux, 2009.

8. Churchland, Patricia Smith. "How Do Neurons Know?" *Daedalus*. Winter 2004: 42–50.

9. Gobet, F., and H. Simon. "Expert Chess Memory: Revisiting the Chunking Hypothesis." *Memory* 6 (1998): 225–255.

10. Gladwell, Malcolm. "The Physical Genius." *The New Yorker*. August 2, 1999.

11. Johnson, Dominic D., and Elizabeth M. P. Madin. "Paradigm Shifts in Security Strategy: Why Does It Take Disasters to Trigger Change?" Pp. 159–185 in Raphael Sagarin and Taylor Terence, eds., *Natural Security: A Darwinian Approach to a Dangerous World*. Berkeley: University of California Press, 2008.

12. Kim, Daniel H. "The Link Between Individual and Organizational Learning." *Sloan Management Review* 35, no. 1 (1993): 37–50.

13. Garvin, D. A., A. C. Edmondson, and F. Gino. "Is Yours a Learning Organization?" *Harvard Business Review* 86, no. 3 (2008): 109ff.

14. Wears, R. L. "Still Learning to Learn." *Quality and Safety in Health Care* 12 (2003): 470–471.

15. Arreguin-Toft, Ivan. "How the Weak Win Wars: A Theory of Asymmetric Conflict." *International Security* 26, no. 1 (2001): 93–128.

16. Krueger, A. B. *The National Origins of Foreign Fighters in Iraq*. 2006. Available at *http://www.aeaweb.org/annual_mtg_papers/2007/0105_1430_1601.pdf*. Accessed July 14, 2010.

17. Vermeij, *Nature: An Economic History*.

18. The White House. "The Federal Response to Hurricane Katrina: Lessons Learned." Washington, DC, 2006.

19. Media Matters. "Fox News Contributor Mike Huckabee Falsely Claimed 'Not One Drop of Oil Was Spilled' During Hurricane Katrina." July 27, 2008. *http://mediamatters.org/research/200806270005*. Accessed August 29, 2010.

20. Broad, William. "Taking Lessons from What Went Wrong." *New York Times*, July 19, 2010.

21. Garvin, D. A. "Building a Learning Organization." *Harvard Business Review* 71, no. 4 (1993): 78–91.

22. Ibid.

23. Weardon, Graeme. "BP's Deepwater Horizon Costs Hit $1.25bn" Guardian. June 7, 2010. *http://www.guardian.co.uk/business/2010/jun/07/bp-deepwater-horizon-costs-soar*. Accessed August 30, 2010.

24. Johnson, Dominic D., and Elizabeth M. P. Madin. "Paradigm Shifts in Security Strategy: Why Does It Take Disasters to Trigger Change?" Pp. 159–185 in Raphael Sagarin and Taylor Terence, eds., *Natural Security: A Darwinian Approach to a Dangerous World*. Berkeley: University of California Press, 2008.

25. March, J. G., L. S. Sproull, and M. Tamuz. "Learning from Samples of One or Fewer." *Organizational Science* 2 (1991): 1–13.

26. Barlow, Nora. *The Autobiography of Charles Darwin: 1809–1882*. New York: W. W. Norton, 1958.

27. March, Sproull, and Tamuz, "Learning from Samples of One or Fewer."

28. Attenborough, David. *Amazing Rare Things: The Art of Natural History in the Age of Discovery*. New Haven, CT: Yale University Press, 2007.

29. Gopnik, Alison. *The Philosophical Baby. What Children's Minds Tell Us About Truth, Love, and the Meaning of Life*. New York: Farrar, Straus, and Giroux, 2009.

30. National Commission on Terrorist Attacks Upon the United States. "The 9/11 Commission Report." Washington, DC, 2004. Several

major news organizations and private blogs from different political persuasions used the "failure of imagination" line in their headlines or ledes in coverage following the release of the report, including: National Public Radio, *http://www.npr.org/911hearings/*; Perrspectives, *http://www.perrspectives.com/blog/archives/000010.htm*; Radio Free Europe Radio Liberty, *http://www.rferl.org/content/article/1053987.html*; the Christian Science Monitor, *http://www.csmonitor.com/2004/0723/p01s03-uspo.html*; and CNN.com, *http://www.cnn.com/2004/ALLPOLITICS/07/22/911.report/index.html*. All accessed August 29, 2010.

CHAPTER FOUR

1. Vermeij, G. J. *Evolution and Escalation: An Ecological History of Life*. Princeton, NJ: Princeton University Press, 1987.

2. Churchland, Patricia Smith. "How Do Neurons Know?" *Daedalus*. Winter 2004: 42–50.

3. Kai Hakkarainen, Kirsti Lonka, and Sami Paavola. 2004. "Networked Intelligence: How Can Human Intelligence Be Augmented Through Artifacts, Communities, and Networks?" Draft paper. *http://www.lime.ki.se/uploads/images/517/Hakkarainen_Lonka_Paavola.pdf*. Accessed August 31, 2010.

4. Vermeij, G. *Nature: An Economic History*, Princeton, NJ: Princeton University Press, 2004, pp. 140–141. Gatesy, S. M., and K. P. Dial. "Locomotor Modules and the Evolution of Avian Flight." *Evolution* 50 (1996): 331–340.

5. Levin, Simon A. *Fragile Dominion*. Cambridge, MA: Perseus, 1999.

6. In Tamm, E. E. *Beyond the Outer Shores: The Untold Odyssey of Ed Ricketts, the Pioneering Ecologist Who Inspired John Steinbeck and Joseph Campbell*. New York: Thunder's Mouth Press, 2004, p. 256.

7. Schmitt, Eric, and David Johnston. "States Chafing at U.S. Focus on Terrorism." *New York Times*, May 26, 2008.

8. Personal comment from the Department of Transportation Maritime Authority, March 2009.

9. Cullins, Douglas. Operational Adaptation Conference. Edinburgh, UK. Personal comments to author on June 23, 2010.

10. Hawken, P. *Blessed Unrest: How the Largest Movement in the World Came into Being and Why No One Saw It Coming.* New York: Viking, 2007.

11. Ibid., pp. 141–144.

12. Brafman, O., and R. A. Beckstrom. *The Starfish and the Spider: The Unstoppable Power of Leaderless Organizations.* New York: Penguin Group, 2006.

13. *http://www.gvfi.org/*. Accessed August 26, 2010.

14. Geddes, Linda. "Disease Maps Can Turn a Crisis Around." *NewScientist,* March 21, 2009, pp. 16–17.

15. Iyer, Bala, and Thomas H. Davenport. "Reverse Engineering Google's Innovation Machine." *Harvard Business Review,* April 1, 2008, pp. 58–68.

16. Ginsberg, J., M. H. Mohebbi, R. S. Patel, L. Brammer, M. S. Smolinski, and L. Brilliant. "Detecting Influenza Epidemics Using Search Engine Query Data." *Nature* 457, no. 7232 (2009): 1012-U4.

17. Glasheen, Jeff. "Spanish Flu Redux." The Hospitalist. October 2009. *http://www.the-hospitalist.org/details/article/366815/Spanish_Flu_Redux.html.* "Pandemic Influenza Awareness Week, Day 1: History of Pandemic Influenza." Aetiology. October 3, 2005. *http://aetiology.blogspot.com/2005/10/pandemic-influenza-awareness-week-day.html.* Both accessed August 26, 2010.

18. A bit of Internet research attributes this widely circulated quote to either mathematician Robert Low or physicist Richard Feynman. See *http://frictionary.blogspot.com/2006_04_01_archive.html* or *http://www.uptoolate.com/rick/memetic/archives/000152.html.* Both accessed April 19, 2010.

19. Trivedi, Bijal. "Thought Control." *NewScientist,* March 1, 2008, pp. 44–47.

20. Krebs, Valdis E. "Mapping Networks of Terrorist Cells." *Connections* 24, no. 3 (2002): 43–52.

21. Sageman, M. "A Strategy for Fighting International Islamist Terrorists." *Annals of the American Academy of Political and Social Science* 618 (2008): 223–231.

22. Gertler, Jeremiah. "V-22 Osprey Tilt-Rotor Aircraft: Background and Issues for Congress." Congressional Research Service Report 7–5700. December 22, 2009.

23. "V-22 Osprey Tilt-Rotor Aircraft Congressional Research Service Report for Congress." Christopher Bolkcom, Congressional Research Service, Library of Congress. Updated January 7, 2005.

24. "Osprey Deployment a Learning Experience." *http://www.navytimes.com/news/2009/08/marine_osprey_082109w/*. Accessed April 13, 2010.

25. Walton, Marsha. "Robots Fail to Complete Grand Challenge: $1 Million Prize Goes Unclaimed." CNN, May 6, 2004. *http://www.cnn.com/2004/TECH/ptech/03/14/darpa.race/index.html*. Boyle, Alan. "Rough Ride for Robots, but Humans Smiling." Msnbc.com, March 14, 2004. *http://www.msnbc.msn.com/id/4517001*. Both accessed April 27, 2010.

26. *http://www.smartgear.org/*. Accessed April 19, 2010.

27. Bowman, Zach. "Edmunds Announces Official Rules for $1 Million Sudden Acceleration Contest." Edmunds.com. April 6, 2010. *http://www.autoblog.com/2010/04/06/edmunds-announces-official-rules-for-1-million-sudden-accelerat/?icid=main|htmlws-main-w|dl6|link4|http%3A%2F%2Fwww.autoblog.com%2F2010%2F04%2F06%2Fedmunds-announces-official-rules-for-1-million-sudden-accelerat%2F*. Accessed August 30, 2010.

28. "Offbeat Ideas for Cleaning the Oil Spill." PRI.org. May 27, 2010. *http://www.pri.org/business/social-entrepreneurs/off-beat-ideas-for-cleaning-the-oil-spill2015.html*. Accessed August 30, 2010.

29. Chaudhuri, Saabira. "250,000-Strong Facebook Group Finds Owner of Lost Camera." Livemint.com. November 4, 2009. *http://www*

.livemint.com/2009/11/04001331/250000-strong-Facebook-group.html?h=B; and Ifoundyourcamera.net. *http://www.ifoundyourcamera.net/*. Both accessed August 30, 2010.

30. Thompson, Mark. "V-22 Osprey: A Flying Shame." *Time*, September 26, 2007.

CHAPTER FIVE

1. Vermeij, G. *Nature: An Economic History*. Princeton, NJ: Princeton University Press, 2004.

2. Arreguin-Toft, Ivan. "How the Weak Win Wars: A Theory of Asymmetric Conflict." *International Security* 26, no. 1 (2001): 93–128.

3. Wasson, Kerstin, and Bruce E. Lyon. "Flight or Fight: Flexible Antipredatory Strategies in Porcelain Crabs." *Behavioral Ecology* 16, no. 6 (2005): 1037–1041.

4. Arquilla, John. "The Coming Swarm." Sunday Opinion, *New York Times*, February 15, 2009.

5. Krakauer, D. C., and V. A. Jansen. "Red Queen Dynamics of Protein Translation. *Journal of Theoretical Biology* 218 (2002): 97–109.

6. Hutchinson, G. E. "An Homage to Santa Rosalia, or Why Are There So Many Kinds of Animals?" *American Naturalist* 93, no. 870 (1959): 145–159. The quote may be apocryphal.

7. Meyer, Axel, and Yves Van de Peer. "'Natural Selection Merely Modified While Redundancy Created: Susumu Ohno's Idea of the Evolutionary Importance of Gene and Genome Duplications." *Journal of Structural and Functional Genomics* 3 (2003): vii–ix.

8. Holmes, Bob. "Accidental Origins." *NewScientist*, March 13, 2010, pp. 30–33; and LePage, "Gene machine," pp. 44–47.

9. Steinbeck, J., and E. F. Ricketts. *Sea of Cortez: A Leisurely Journal of Travel and Research, with a Scientific Appendix Comprising Materials for a Source Book on the Marine Animals of the Panamic Faunal Province*. Mamaroneck, NY: P. P. Appel, (1941) 1971.

10. Erlich, Paul, and Anne Erlich. *Extinction: The Causes and Consequences of the Disappearance of Species.* New York: Random House, 1981.

11. Kish, Daniel. "Seeing with Sound." *NewScientist,* April 11, 2009, pp. 31–33.

12. Asch, Peter, Burton G. Malkiel, and Richard E. Quandt. "Racetrack Betting and Informed Behavior." *Journal of Financial Economics* 10, no. 2 (1982): 187–194. There is some variation in the success of different bettors, however.

13. Wikipedia. "Futures exchange." *http://en.wikipedia.org/wiki/Futures_exchange.* Accessed April 20, 2010.

14. Chittka L., P. Skorupski, and N. E. Raine. "Speed-Accuracy Tradeoffs in Animal Decision Making." *Trends in Ecology & Evolution* 24 (2009): 400–407.

15. "Crowdsourcing the Crystal Ball." *http://www.forbes.com/2007/10/13/james-surowiecki-prediction-tech-future07-cx_js_1015wisdom.html.* Accessed April 20, 2010.

16. Ananthaswamy, Anil. "Mobilising the Minds of the Masses." *NewScientist,* February 14, 2009, pp. 20–21.

17. C. Josh Donlan, Joel Berger, Carl E. Bock, Jane H. Bock, David A. Burney, James A. Estes, Dave Foreman, Paul S. Martin, Gary W. Roemer, Felisa A. Smith, Michael E. Soulé, and Harry W. Greene. "Pleistocene Rewilding: An Optimistic Agenda for Twenty-First Century Conservation." *American Naturalist* 168, no. 5 (2006).

18. James T. Mandel, C. Josh Donlan, and Jonathan Armstrong. "A Derivative Approach to Endangered Species Conservation." *Frontiers in Ecology and the Environment* 8, no. 1 (2010): 44–49. doi: 10.1890/070170.

19. "Regulators Approve Movie Box Office Futures Market." *http://www.huffingtonpost.com/2010/04/16/regulators-approve-movie-_n_541085.html.* Accessed April 20, 2010.

20. "New Box Office Futures Market Could Allow You to Bet on

Movies." *http://www.switched.com/2010/04/19/new-box-office-futures-market-could-allow-you-to-bet-on-movies/*. Accessed April 20, 2010.

21. "Amid Furor, Pentagon Kills Terrorism Futures Market." *http://www.cnn.com/2003/ALLPOLITICS/07/29/terror.market/*. Accessed April 20, 2010.

22. "The Case for Terrorism Futures." *http://www.wired.com/politics/law/news/2003/07/59818*. Accessed April 20, 2010.

23. Vespignani, A. "Complex Networks: The Fragility of Interdependency." *Nature* 464 (2010): 984–985.

24. Berlow, Eric. "How Complexity Leads to Simplicity." TED talks. July 2010. Oxford, England. *http://blog.ted.com/2010/11/12/how-complexity-leads-to-simplicity-eric-berlow-on-ted-com/*. Accessed April 15, 2011.

25. Barabasi, A.-L. *Linked*. London: Penguin, 2003.

26. Sageman, M. "A Strategy for Fighting International Islamist Terrorists." *Annals of the American Academy of Political and Social Science* 618 (2008): 223–231. Atran, Scott. Statement Before the Senate Armed Services Subcommittee on Emerging Threats & Capabilities, "Pathways to and From Violent Extremism: The Case for Science-Based Field Research." March 10, 2010.

27. Krebs, Valdis E. "Mapping Networks of Terrorist Cells." *Connections* 24 (2002): 43–52.

28. Jordan, F. "Network Analysis Links Parts to the Whole." Pp. 240–260 in Raphael Sagarin and Terence Taylor, eds., *Natural Security: A Darwinian Approach to a Dangerous World*. Berkeley: University of California Press, 2008.

29. Buldyrev, S. V., R. Parshani, G. Paul, H. E. Stanley, and S. Havlin. "Catastrophic Cascade of Failures in Interdependent Networks." *Nature* 464 (2010): 1025–1028; and Vespignani, "Complex Networks: The Fragility of Interdependency."

30. Vespignani, "Complex Networks: The Fragility of Interdependency."

CHAPTER SIX

1. *http://www.imdb.com/title/tt0086567/quotes.* Accessed July 30, 2010.

2. Scheffer, Marten. "Alternative Stable States and Regime Shifts in Ecosystems." Pp. 395–406 in Simon A. Levin, ed., *The Princeton Guide to Ecology,* Princeton, NJ: Princeton University Press, 2010.

3. *http://www.suddenoakdeath.org/html/history___background.html.* Accessed August 2, 2010.

4. Biosphere2. *http://www.b2science.org/b2/about-history.html.* Accessed August 2, 2010.

5. Alvarez, Luis W., Walter Alvarez, Frank Asaro, and Helen V. Michel. "Extraterrestrial Cause for the Cretaceous-Tertiary Extinction: Experimental Results and Theoretical Interpretation." *Science* 208 (1980): 4448.

6. "Cretaceous tertiary extinction event." *http://en.wikipedia.org/wiki/Cretaceous%E2%80%93Tertiary_extinction_event#Duration.* Accessed August 2, 2010.

7. Turco, R. P., O. B. Toon, T. P. Ackerman, J. B. Pollack, and C. Sagan. "Nuclear Winter: Global Consequences of Multiple Nuclear Explosions." *Science* 222, no. 4630 (1983): 1283–1292.

8. Warden, John K. "Ambassador Linton Brooks on New START and the Next Treaty." April 16, 2010. *http://csis.org/blog/ambassador-linton-brooks-new-start-and-next-treaty*; Oelrich, Ivan. "New START and Missile Defense." July 27, 2010. *http://www.fas.org/blog/ssp/2010/07/new-start-and-missile-defense.php.*; and Kaplan, Fred. "Reagan's Nuclear Defense Strategy: Myth and Reality." Cato Institute Policy Analysis. January 30, 1982. *http://www.cato.org/pubs/pas/pa006.html.* All accessed August 9, 2010.

9. O'Donoghue, James. "The Second Coming." *NewScientist,* June 14, 2008.

10. Vermeij, Geerat J. *Evolution and Escalation: An Ecological History of Life.* Princeton, NJ: Princeton University Press, 1987.

11. Personal communication with Air National Guard Major Noel Lipana, September 15, 2009.

12. Eisler, P. "Insurgents Adapt Faster Than Military Adjusts to IEDs." *USA Today*, July 16, 2007.

13. "COIN Adaptation in Afghanistan and Iraq." Presentation at the Conference on Organizational Adaptation. Edinburgh, Scotland. June 24, 2010.

14. Weiner, Jonathan. *The Beak of the Finch*. New York: Alfred A. Knopf, 1994.

15. "Instant Evolution Helps Make Up for Vanishing Pollinators." *NewScientist*, July 10, 2010.

16. Kenney, Michael. *From Pablo to Osama: Trafficking and Terrorist Networks, Government Bureaucracies, and Competitive Adaptation*. University Park: Pennsylvania State University Press, 2007. P. 104.

17. Neuman, William, and Andrew Pollack. "Farmers Cope with Roundup-Resistant Weeds." *New York Times*, May 3, 2010.

18. Romano, Jay. "Bedbugs." *New York Times*, September 27, 2006. *http://topics.nytimes.com/top/reference/timestopics/subjects/b/bedbugs/index.html*. Accessed August 10, 2010. Berenbaum, May. "This Bedbug's Life." *New York Times*, August 7, 2010.

19. Mayo Clinic Staff. "MRSA Infection." *http://www.mayoclinic.com/health/mrsa/DS00735*. Accessed August 10, 2010. "Methicillin-resistant Staphylococcus aureus." *http://en.wikipedia.org/wiki/Methicillin-resistant_Staphylococcus_aureus*. Accessed August 10, 2010.

20. National Commission on Terrorist Attacks Upon the United States. "The 9/11 Commission Report." Washington, DC, 2004.

21. Dawkins, R., and J. R. Krebs. "Arms Races Between and Within Species." *Proceeding of the Royal Society B: Biological Sciences* 205, no. 1161 (1979): 489–511.

22. Kenney, Michael. *From Pablo to Osama: Trafficking and Terrorist Networks, Government Bureaucracies, and Competitive Adaptation*. University Park: Pennsylvania State University Press, 2007. P. 118.

CHAPTER SEVEN

1. *www.marmotburrow.ucla.edu.*

2. Atran, Scott. Statement Before the Senate Armed Services Subcommittee on Emerging Threats & Capabilities. "Pathways to and from Violent Extremism: The Case for Science-Based Field Research." March 10, 2010.

3. Kenney, Michael. *From Pablo to Osama: Trafficking and Terrorist Networks, Government Bureaucracies, and Competitive Adaptation.* University Park: Pennsylvania State University Press, 2007. "Traffickers and terrorists enjoy an important advantage over their government opponents: they know when, where and how they will carry out their activities: law enforcers do not." P. 204.

4. Ibid., p. 207.

5. Atran, Scott. Statement Before the Senate Armed Services Subcommittee on Emerging Threats & Capabilities. "Pathways to and from Violent Extremism: The Case for Science-Based Field Research." March 10, 2010.

6. Barr, Andy. "Rep. Mike Rogers: Execute WikiLeaks Leaker." Politico. August 3, 2010. *http://www.politico.com/news/stories/0810/40599.html.* Accessed August 4, 2010.

7. Brooks, Michael. "Do You Speak Cuttlefish?" *NewScientist,* April 26, 2008. Mathger, L. M., E. J. Denton, N. J. Marshall, and R. T. Hanlon. "Mechanisms and Behavioural Functions of Structural Coloration in Cephalopods." *Journal of the Royal Society Interface* 6 (2009): S149-S63.

8. Personal communication from Air National Guard Major Noel Lipana, September 15, 2009.

9. Kenney, Michael. *From Pablo to Osama: Trafficking and Terrorist Networks, Government Bureaucracies, and Competitive Adaptation.* University Park: Pennsylvania State University Press, 2007.

10. Geddes, Linda. "How Consistent Criminals Give Themselves Away." *NewScientist,* August 9, 2008.

11. Birch, Hayley. "Bugging Your Bugs." *NewScientist,* March 6, 2010.

12. Webb, G. F. "The Prime Number Periodical Cicada Problem." *Discrete and Continuous Dynamical Systems, Series B* 1, no. 3 (2001): 387–399.

13. Department of Homeland Security. "Chronology of Changes to the Homeland Security Advisory System." *http://www.dhs.gov/xabout/history/editorial_0844.shtm*. Accessed August 9, 2010.

14. Lehrer, Jonah. "Under Pressure: The Search for a Stress Vaccine." *Wired*, July 28, 2010. Sapolsky, Robert. *A Primate's Memoir.* New York: Scribner, 2001.

15. Tierney, John. "Living in Fear and Paying a High Cost in Heart Risk." *New York Times*, January 15, 2008. Holman, E. A., R. C. Silver, M. Poulin, J. Andersen, V. Gil-Rivas, and D. N. McIntosh. "Terrorism, Acute Stress, and Cardiovascular Health." *Archives of General Psychiatry* 65, no. 1 (2008): 73–80.

16. Chittka, L., P. Skorupski, and N. E. Raine. "Speed-Accuracy Tradeoffs in Animal Decision Making." *Trends in Ecology & Evolution* 24, no. 7 (2009): 400–407.

17. "Oops, Sorry, I Thought You Were Someone Else." *NewScientist*, April 26, 2008.

18. Brooks, Michael. "Do You Speak Cuttlefish?" *NewScientist*, April 26, 2008.

19. BBC News. "Two Escape from an Argentine Jail Guarded by a Dummy." July 20, 2010. *http://www.bbc.co.uk/news/world-latin-america-10706626*. Accessed August 9, 2010.

20. Brooks, Michael. "Do You Speak Cuttlefish?" *NewScientist*, April 26, 2008.

21. Rundus, Aaron S., Donald H. Owings, Sanjay S. Joshi, Erin Chinn, and Nicholas Giannini. "Ground Squirrels Use an Infrared Signal to Deter Rattlesnake Predation." *Proceedings of the National Academy of Science* 104, no. 36 (2007): 14372–14376.

22. Orr, Bob. "Videos Demystify the Osama bin Laden Legend." May 7, 2011. CBSnews.com. Accessed July 5, 2011.

23. Husain, Ed. "Did U.S. Botch Message with bin Laden Videos?" Council on Foreign Relations. 2011. *http://www.cfr.org/terrorism/did-us-botch-message-bin-laden-videos/p24939.* Accessed July 5, 2011. Scott Atran reports that commentators on radical Muslim websites after the video releases indeed saw the bin Laden videos as indications that he continued to live a devout life, despite his persecution.

24. Jones, R. V. *Most Secret War.* London: Hamilton, 1978.

25. Blumstein, D. T. "Fourteen Security Lessons from Antipredator Behavior." Pp. 147–158 in Raphael Sagarin and Terence Taylor, eds., *Natural Security: A Darwinian Approach to a Dangerous World.* Berkeley: University of California Press, 2008.

26. Ibid.

27. Wald, Matthew L. "For No Signs of Trouble, Kill the Alarm." *New York Times,* August 1, 2010.

28. The Fire Marshal's Public Fire Safety Council. "Fire Alarm Fact Sheet." *http://www.firesafetycouncil.com/english/pubsafet/safact.htm.* United States Fire Administration. "Smoke Alarms." *http://www.usfa.dhs.gov/downloads/pyfff/smkalarm.html.* Both accessed August 11, 2010.

29. Dziekan, Mike. "Where There's Smoke There's (Not Always) Fire: An Inside Look at Smoke Detectors." Society for Amateur Scientists. 2004. *http://www.sas.org/tcs/weeklyIssues/2004–07–30/feature1/index.html.* Accessed August 2, 2010.

30. Chittka, L., P. Skorupski, and N. E. Raine. "Speed-Accuracy Tradeoffs in Animal Decision Making." *Trends in Ecology & Evolution* 24: 400–407.

31. "Fort Hood shooting." *http://en.wikipedia.org/wiki/Fort_Hood_shooting.* Accessed August 11, 2010.

32. Carre, J. M., and C. M. McCormick. "In Your Face: Facial Metrics Predict Aggressive Behaviour in the Laboratory and in Varsity and Professional Hockey Players." *Proceedings of the Royal Society B–Biological Sciences* 275, no. 1651 (2008): 2651–2656.

33. Sell, A., L. Cosmides, J. Tooby, D. Sznycer, C. von Rueden, and M.

Gurven. "Human Adaptations for the Visual Assessment of Strength and Fighting Ability from the Body and Face." *Proceedings of the Royal Society B: Biological Sciences* 276, no. 1656 (2009): 575–584.

34. Tibbetts, E. A., and J. Dale. "Individual Recognition: It Is Good to Be Different." *Trends in Ecology & Evolution* 22, no. 10 (2007): 529–537.

35. Marzluff, John M., Jeff Walls, Heather N. Cornell, John C. Withey, and David P. Craig. "Lasting Recognition of Threatening People by Wild American Crows." *Animal Behaviour* 79, no. 3 (2010): 699–707. Cornell, H. N., J. M. Marzluff, and S. Pecoraro. "Social Learning Spreads Knowledge About Dangerous Humans Among American Crows." *Proceedings of the Royal Society of B: Biological Sciences* (2011).doi: 10.1098 /rspb.2011.0957.

36. Darwin, Charles. *The Expressions of the Emotions in Man and Animals.* London: John Murray, 1872.

37. "Behavior Detection Officers (BDO)." *http://www.tsa.gov/what_we_do/layers/bdo/index.shtm*. Accessed August 10, 2010.

38. Mills, Doug. "Faces, Too, Are Searched at U.S. Airports " *New York Times,* August 17, 2006.

39. Burgoon, J. K., D. P. Twitchell, M. L. Jensen, T. O. Meservy, M. Adkins, J. Kruse, A. V. Deokar, G. Tsechpenakis, S. Lu, D. N. Metaxas, J. F. Nunamaker, and R. E. Younger. "Detecting Concealment of Intent in Transportation Screening: A Proof of Concept." *IEEE Transactions on Intelligent Transportation Systems* 10, no. 1 (2009): 103–112.

40. Government Accountability Office. "Efforts to Validate TSA's Passenger Screening Behavior Detection Program Underway, but Opportunities Exist to Strengthen Validation and Address Operational Challenges." GAO-10–763. May 2010. Weinberger, Sharon. "Intent to Deceive?" *Nature* 466 (2010): 412–415.

41. Yu, Roger. "Airport Check-in: TSA Behavior Screening Misses Suspects." May 24, 2010. *http://travel.usatoday.com/flights/legacy/item.aspx?type=blog&ak=93938.blog.* Transportation Security Administration. "TSA SPOT Program: Still Going Strong." May 21, 2010.

http://blog.tsa.gov/2010/05/tsa-spot-program-still-going-strong.html?showComment=1274464283131. Both accessed August 10, 2010.

42. Government Accountability Office. "Efforts to Validate TSA's Passenger Screening Behavior Detection Program Underway, but Opportunities Exist to Strengthen Validation and Address Operational Challenges."GAO-10–763. May 2010.

43. September 30, 2010, response to August 26, 2010, e-mail request to TSA by the author from Oso-Correspondence@dhs.gov.

44. Boarding data from FAA, available at *http://www.faa.gov/airports/planning_capacity/passenger_allcargo_stats/passenger/*. Accessed August 11, 2010.

CHAPTER EIGHT

1. Much of this work has come out of the Center for Evolutionary Psychology at University of California, Santa Barbara, run by John Tooby and Leda Cosmedies. *http://www.psych.ucsb.edu/research/cep/*. Accessed August 24, 2010.

2. Stancliff, Dave. "History Shows Why We Should Get Out of Afghanistan." *Times Standard*, September 27, 2009. *http://www.times-standard.com/davestancliff/ci_13432321*. Accessed August 25, 2010. A detailed history of different countries' blunders into Afghanistan is in Coll, Steve. *Ghost Wars: The Secret History of the CIA, Afghanistan, and Bin Laden, from the Soviet Invasion to September 10, 2001*. New York: Penguin, 2004.

3. Dawkins, Richard. *The God Delusion*. New York: Houghton Mifflin, 2006. Hitchens, Christopher. *God Is Not Great*. New York: Hachette, 2007.

4. Wilson, Edward. O. *The Creation: An Appeal to Save Life on Earth*. New York: W. W. Norton, 2006.

5. Xatal.com. "Which UCs Are the Most Selective?" November 13, 2009. *http://xatal.com/california/which-ucs-are-the-most-selective/*. Accessed August 25, 2010.

6. Villarreal, Luis P., and Victor R. DeFilippis. "A Hypothesis for DNA Viruses as the Origin of Eukaryotic Replication Proteins." *Journal of Virology* 74, no. 15 (2000): 7079–7084.

Villarreal, Luis P., and Guenther Witzany. "Viruses Are Essential Agents within the Roots and Stem of the Tree of Life." *Journal of Theoretical Biology* (in press).

7. Villarreal, Luis P. "From Biology to Belief." Pp. 42–68 in Raphael Sagarin and Taylor Terence, eds., *Natural Security: A Darwinian Approach to a Dangerous World.* Berkeley: University of California Press, 2008.

8. Den Boer, S. P. A., B. Baer, and J. J. Boomsma. "Seminal Fluid Mediates Ejaculate Competition in Social Insects." *Science* 327, no. 5972 (2010): 1506–1509.

9. Doving, Kjell B., and Didier Trotier. "Structure and Function of the Vomeronasal Organ." *Journal of Experimental Biology* 201 (1998): 2913–2925.

10. Doty, Richard L. "The Pheromone Myth: Sniffing Out the Truth." *NewScientist,* March 1, 2010.

11. Sergeant, M. J. T., T. E. Dickins, M. N. O. Davies, and M. D. Griffiths. "Women's Hedonic Ratings of Body Odor of Heterosexual and Homosexual Men." *Archives of Sexual Behavior* 36, no. 3 (2007): 395–401. This study, for example, provides the nonintuitive finding that heterosexual females prefer the smell of homosexual, rather than heterosexual, men.

12. Ravilious, Kate. "Messages from the Stone Age." *NewScientist,* February 20, 2010, pp. 30–34.

13. Oomoto. "History of Oomoto." *http://www.oomoto.or.jp/English/enHist/histm-en.html.* Accessed August 23, 2010.

14. Henrich, J. "The Evolution of Costly Displays, Cooperation, and Religion: Credibility-Enhancing Displays and Their Implications for Cultural Evolution." *Evolution and Human Behavior* 30, no. 4 (2009): 244–260.

15. Ibid.

16. "The Puzzle of Human Cooperation." *Nature* 421 (2003): 911–912.

17. Atran, Scott, Robert Axelrod, and Richard Davis. "Sacred Barriers to Conflict Resolution." *Science* 317 (2007): 1039–1040.

18. Ibid.

19. Atran, S. "The Power of Moral Belief." Pp. 141–144 in Raphael Sagarin and Taylor Terence, eds., *Natural Security: A Darwinian Approach to a Dangerous World*. Berkeley: University of California Press, 2008.

20. Leonard, Tom. "America's Armed Militia on the Rise." *Telegraph*, December 31, 2009. *http://www.telegraph.co.uk/news/worldnews/northamerica/usa/6917525/Americas-armed-militia-on-the-rise.html*. Accessed August 26, 2010. Southern Poverty Law Center. "The Second Wave: Return of the Militias." Montgomery, AL: SPLC, 2009.

21. Koerner, Brendan I. "Secret of AA: After 75 Years, We Don't Know How It Works." *Wired*, June 23, 2010. *http://www.wired.com/magazine/2010/06/ff_alcoholics_anonymous/all/1*. Accessed August 24, 2010.

22. Sosis, Richard, and Candace S. Alcorta. "Militants and Martyrs: Evolutionary Perspectives on Religion and Terrorism." Pp. 105–124 in Raphael Sagarin and Taylor Terence, eds., *Natural Security: A Darwinian Approach to a Dangerous World*. Berkeley: University of California Press, 2008.

23. Dobbs, David. 2011. "Beautiful Brains." *National Geographic*, October, 2011: 36–59.

24. Atran, Scott. Statement Before the House Appropriations Subcommittee on Homeland Security. "The Making of a Terrorist: A Need for Understanding from the Field." March 12, 2008.

25. Wheelock, Darren, and Douglas Hartmann. "Midnight Basketball and the 1994 Crime Bill Debates: The Operation of a Racial Code." *The Sociological Quarterly* 48 (2007): 315–342.

26. Atran, Scott. Statement Before the Senate Armed Services Subcommittee on Emerging Threats & Capabilities. "Pathways to and from Violent Extremism: The Case for Science-Based Field Research." March 10, 2010.

CHAPTER NINE

1. Eshel, David. "IEDs." *Journal of Electronic Defense* (2007): 39–42.

2. Sachs, Joel L., and Ellen L. Simms. "Pathways to Mutualism Breakdown." *Trends in Ecology & Evolution* 21, no. 10 (2006): 585–592.

3. Brahic, Catherine. "'Pandora' Bacteria Act as One Organism." *NewScientist,* February 27, 2010.

4. Herrera, C. M., and M. I. Pozo. "Nectar Yeasts Warm the Flowers of a Winter-Blooming Plant." *Proceedings of the Royal Society B–Biological Sciences* 277, no. 1689 (2010): 1827–1834.

5. Margulis, Lynn. *Symbiotic Planet: A New Look at Evolution.* New York: Basic, 1998.

6. Haddock, Steven H. D., Mark A. Moline, and James F. Case. "Bioluminescence in the Sea." *Annual Review of Marine Science* 2 (2010): 443–493.

7. Lefevre, T., C. Lebarbenchon, M. Gauthier-Clerc, D. Misse, R. Poulin, and F. Thomas. "The Ecological Significance of Manipulative Parasites." *Trends in Ecology & Evolution* 24, no. 1 (2009): 41–48.

8. MacKenzie, Debora. "Parasite Lost" *NewScientist,* March 13, 2010, pp. 39–41.

9. Horgan, John. "The End of War." *NewScientist,* July 4, 2009, pp. 38–41.

10. Cheney, K. L., R. Bshary, and A. S. Grutter. "Cleaner Fish Cause Predators to Reduce Aggression Toward Bystanders at Cleaning Stations." *Behavioral Ecology* 19, no. 5 (2008): 1063–1067.

11. Allee, W. C. *Cooperation Among Animals with Human Implications.* New York: Henry Schuman, 1951.

12. Allee, W. C. "Where Angels Fear to Tread: A Contribution from General Sociology to Human Ethics." *Science* 97, no. 2528 (1943): 517–525.

13. Ibid.

14. Allee, W. C. *Cooperation Among Animals with Human Implications.* New York: Henry Schuman, 1951, p. 195.

15. Ibid., p. 197.

16. Ibid., p. 199.

17. Rischard, Jean-François. "Global Issues Networks: Desperate Times Deserve Innovative Measures." *Washington Quarterly* (Winter 2002-2003): 17–33.

18. Liebhold, Andrew M. "Warder Allee's Escape from Obscurity." *Trends in Ecology & Evolution* 23, no. 6 (2008): 297–298.

19. "Warder Clyde Allee." *http://en.wikipedia.org/wiki/Warder_Clyde_Allee*. Accessed August 19, 2010.

20. Margulis, Lynn. *Symbiotic Planet: A New Look at Evolution*. New York: Basic, 1998.

21. Ibid., p. 37.

22. Svensson, E. I. "Understanding the Egalitarian Revolution in Human Social Evolution." *Trends in Ecology & Evolution* 24, no. 5 (2009): 233–235; Sigmund, K. "Punish or Perish? Retaliation and Collaboration Among Humans." *Trends in Ecology & Evolution* 22, no. 11 (2007): 593–600.

23. Ratnieks, Francis L. W., and Tom Wenseleers. "Altruism in Insect Societies and Beyond: Voluntary or Enforced?" *Trends in Ecology & Evolution* 23, no. 1 (2007): 45–52.

24. Choi, J. K., and S. Bowles. "The Coevolution of Parochial Altruism and War." *Science* 318, no. 5850 (2007): 636–640.

25. Gavrilets, S., E. A. Duenez-Guzman, and M. D. Vose. "Dynamics of Alliance Formation and the Egalitarian Revolution." *PLOS One* 3, no. 10 (2008); Svensson, E. I. "Understanding the Egalitarian Revolution in Human Social Evolution." *Trends in Ecology & Evolution* 24, no. 5 (2009): 233–235.

26. De Quervain, D. J. F., U. Fischbacher, V. Treyer, M. Schellhammer, U. Schnyder, A. Buck, and E. Fehr. "The Neural Basis of Altruistic Punishment." *Science* 305, no. 5688 (2004): 1254–1258.

27. Wallace, B., D. Cesarini, P. Lichtenstein, and M. Johannesson. "Heritability of Ultimatum Game Responder Behavior." *Proceedings of*

the National Academy of Sciences of the United States of America 104, no. 40 (2007): 15631–15634.

28. Axelrod, Robert. "The Evolution of Cooperation." New York: Basic, 1984.

29. Sigmund, K. "Punish or Perish? Retaliation and Collaboration Among Humans." *Trends in Ecology & Evolution* 22, no. 11 (2007): 593–600.

30. Leventhal, A., A. Ramlawi, A. Belbiesi, and R. D. Balicer. "Regional Collaboration in the Middle East to Deal with H5N1 Avian Flu." *British Medical Journal* 333 (2006): 856–858.

31. Gresham, Louise, Assad Ramlawi, Julie Briski, Mariah Richardson, and Terence Taylor. "Trust Across Borders: Responding to 2009 H1n1 Influenza in the Middle East." *Biosecurity and Bioterrorism: Biodefense Strategy, Practice, and Science* 7, no. 4 (2009): 399–404.

32. Leventhal, A., A. Ramlawi, A. Belbiesi, and R. D. Balicer. "Regional Collaboration in the Middle East to Deal with H5N1 Avian Flu." *British Medical Journal* 333 (2006): 856–858.

33. Fischhoff, Baruch, Scott Atran, and Marc Sageman. "Mutually Assured Support: A Security Doctrine for Terrorist Nuclear Weapon Threats." *Annals of the American Academy of Political and Social Science* 618 (2008): 160–167.

34. "Trench warfare." *http://www.wordiq.com/definition/Trench_warfare*. Accessed August 12, 2010.

35. The Carter Center. "Historic Cease-Fire Enables Health Workers to Attack Guinea Worm and Other Diseases in Sudan." June 17, 1995. *http://www.cartercenter.org/news/documents/doc169.html*. Accessed August 12, 2010.

36. Bronstein, Judith L. "Mutualism and Symbiosis." Pp. 233–238 in Simon Levin, ed., *The Princeton Guide to Ecology*. Princeton, NJ: Princeton University Press, 2009.

37. Kiers, E. T., R. A. Rousseau, S. A. West, and R. F. Denison. "Host Sanctions and the Legume-Rhizobium Mutualism." *Nature* 425, no. 6953 (2003): 78–81.

38. Bronstein, Judith L. "Mutualism and Symbiosis." Pp. 233–238 in Simon Levin, ed., *The Princeton Guide to Ecology*. Princeton, NJ: Princeton University Press, 2009.

39. Personal communication from Richard Cudney Bueno, September 2010.

40. Allee, W. C. "Where Angels Fear to Tread: A Contribution from General Sociology to Human Ethics." *Science* 97, no. 2528 (1943): 517–525.

CHAPTER TEN

1. McDougall, C. *Born to Run: A Hidden Tribe, Superathletes, and the Greatest Race the World Has Never Seen*. New York: Alfred A. Knopf, 2009, p. 176.

2. Ibid., p. 177.

3. Matthews, Robert. "The Heat Is On." *NewScientist*, July 31, 2010, pp. 42–45.

4. The Interagency Task Force on Antimicrobial Resistance and *A Public Health Action Plan to Combat Antimicrobial Resistance. http://www.cdc.gov/drugresistance/actionplan/index.htm*. Facts About Antibiotic Resistance. *http://www.idsociety.org/Content.aspx?id=5650*. Both accessed May 6, 2010.

5. Nesse, R. M., and G. C. Williams. *Why We Get Sick: The New Science of Darwinian Medicine*. New York: Vintage, 1994.

6. Nesse, R. M., and S. C. Stearns. "The Great Opportunity: Evolutionary Applications to Medicine and Public Health." *Evolutionary Applications* 1 (2008): 28–48.

7. Ostfeld, R. S., and F. Keesing. "Biodiversity and Disease Risk: The case of Lyme Disease." *Conservation Biology* 14 (2000): 722–728.

8. Dobson, A., et al. "Habitat Loss, Trophic Collapse, and the Decline of Ecosystem Services." *Ecology* 87 (2006): 1915–1924.

9. Ibid. In addition, see Keesing, F., R. D. Holt, and R. S. Ostfeld. "Effects of Species Diversity on Disease Risk." *Ecology Letters* 9 (2006):

485–498; and Pongsiri, M. J., J. Roman, V. O. Ezenwa, T. L. Goldberg, H. S. Koren, S. C. Newbold, R. S. Ostfeld, S. K. Pattanayak, and D. J. Salkeld. "Biodiversity Loss Affects Global Disease Ecology." *BioScience* 59 (2009): 945–954. These are all good reviews of the relationship between biodiversity and infectious disease.

10. United Nations. *Human Development Report 2006: Beyond Scarcity—Power, Poverty, and the Global Water Crisis.* New York: United Nations Development Programme, 2006.

11. Buttice, A. L., J. M. Stroot, D. V. Lim, P. G. Stroot, and N. A. Alcantar. "Removal of Sediment and Bacteria from Water Using Green Chemistry." *Environmental Science and Technology* 44 (2010): 3514–3519.

12. Davis, M. *Ecology of Fear: Los Angeles and the Imagination of Disaster.* New York: Vintage, 1999.

13. Gumprecht, B. *The Los Angeles River: Its Life, Death, and Possible Rebirth.* Baltimore: Johns Hopkins University Press, 2001.

14. Ibid.

15. Friends of Vast Industrial Concrete Kafkaesque Structures. *http://seriss.com/people/erco/fovicks/*. Accessed April 29, 2010.

16. The White House. "The Federal Response to Hurricane Katrina: Lessons Learned." Washington, DC, 2006.

17. Day, J. W., et al. "Restoration of the Mississippi Delta: Lessons from Hurricanes Katrina and Rita." *Science* 315 (2007): 1679–1684.

18. The White House. "The Federal Response to Hurricane Katrina: Lessons Learned." Washington, DC, 2006.

19. Day, J. W., et al. "Restoration of the Mississippi Delta: Lessons from Hurricanes Katrina and Rita." *Science* 315 (2007): 1679–1684.

20. Granek, E. F., and B. Ruttenberg. "The Protective Capacity of Mangroves During Tropical Storms: A Case Study from Wilma and Gamma in Belize." Western Society of Naturalists Annual Meeting. Ventura, CA, 2007.

21. "Mangroves Shielded Communities Against Tsunami." October 28, 2005. *http://www.sciencedaily.com/releases/2005/10/051028141252.htm.*

Accessed May 7, 2010. Danielsen, F., et al. "The Asian Tsunami: A Protective Role for Coastal Vegetation." *Science* 310 (2005): 643.

22. The White House. "The Federal Response to Hurricane Katrina: Lessons Learned." Washington, DC, 2006.

23. "Preparing the Ike Dike Defense." *Wall Street Journal,* June 4, 2009. *http://online.wsj.com/article_email/SB124407051124382899-lMyQjAxMDI5NDE0OTAxNzkwWj.html.* Accessed May 7, 2010.

24. "Migrants Finding Ways to Climb 18-Foot-Tall Border Fence." *Arizona Republic,* November 15, 2008. *http://www.tucsoncitizen.com/ss/related/102700.* Accessed June 3, 2010.

25. "Napolitano Bludgeons Border-Fence Proposal." WorldNetDaily.com. December 21, 2005. *http://www.wnd.com/?pageId=34017.* Accessed August 10, 2010. Also see ibid.

26. Wulf, William. Cyber Security Panel. Institute on Science for Global Policy Conference. Tucson, AZ, December 8, 2009.

27. Wulf, William A., and Anita K. Jones. "Reflections on Cybersecurity." *Science* 326 (2009): 943–944.

28. Gordon, Deborah. *Ant Encounters: Interaction Networks and Colony Behavior.* Princeton, NJ: Princeton University Press, 2010.

29. Industrial Control Systems Cyber Emergency Response Team. "USB Drives Commonly Used as an Attack Vector Against Critical Infrastructure." 2010.

30. "Recommendations for Appropriate Shoreline Stabilization Methods for the Different North Carolina Estuarine Shoreline Types." North Carolina Estuarine Biological and Physical Processes Work Group and North Carolina Division of Coastal Management, 2006.

31. Dorsey, J., P. M. Carter, S. Bergquist, and R. Sagarin. "Reduction of Fecal Indicator Bacteria (FIB) in the Ballona Wetlands Salt-water Marsh (Los Angeles County, California, USA) with Implications for Restoration Actions." *Water Research* 44, no. 15 (2010): 4630–4642.

32. Margulis, Lynn. *Symbiotic Planet: A New Look at Evolution*. New York: Basic, 1998.

33. Daily, Gretchen, ed. *Nature's Services: Societal Dependence on Natural Ecosystems*. Washington, DC: Island Press, 1997.

34. "High Economic Value Set on Threatened Mexican Mangroves." *Science Daily,* July 23, 2008. *http://www.sciencedaily.com/releases/2008/07/080721173757.htm*. Accessed May 4, 2010.

CONCLUSION

1. "Pollution Prevention Pays." *http://solutions.3m.com/wps/portal/3M/en_US/3M-Sustainability/Global/Environment/3P/*. Accessed May 16, 2011.

2. Sobel, Dava. *Longitude: The True Story of a Lone Genius Who Solved the Greatest Scientific Problem of His Time*. New York: Penguin, 1996.

3. Lincicome, Bernie. "Maybe Sitting Bull Had A Power Forward." *Chicago Tribune,* June 12, 1996.

4. Skeets, J. E. "Phil Jackson: Rodman Is the Greatest Athlete I've Ever Coached." Sports.yahoo.com.

5. "Online gamers enlisted by University of Washington deliver big-time scientific results." Published: Sunday, September 18, 2011, 10:00 AM. Updated: Wednesday, September 21, 2011, 3:33 PM by Joe Rojas-Burke, *The Oregonian.*

6. Turnipseed, M., R. Sagarin, et al. "Reinvigorating the Public Trust Doctrine: Expert Opinion on the Potential of a Public Trust Mandate in U.S. and International Environment Law." *Environment* 52, no. 5 (2010): 6–14.

7. Turnipseed, M., S. E. Roady, et al. "The Silver Anniversary of the United States' Exclusive Economic Zone: Twenty-Five Years of Ocean Use and Abuse, and the Possibility of a Blue Water Public Trust Doctrine." *Ecology Law Quarterly* 36, no. 1 (2009): 1–70.

8. Barnes, Peter. *Who Owns the Sky? Our Common Assets and the Future of Capitalism.* Washington, DC: Island Press, 2001.

9. Foster, Mary, and Holbrook Mohr. "La. Spillway to Open, Flooding Cajun Country." Associated Press, May 13, 2011. *http://news.yahoo.com/s/ap/20110513/ap_on_re_us/us_mississippi_river_flooding.* Accessed May 13, 2011.

INDEX

J

K

CPSIA information can be obtained
at www.ICGtesting.com
Printed in the USA
JSHW041405190921
18770JS00003B/169

9 780465 021833